Digital Lives in

the Global City

Digital
LIVES
in the
Global City

Contesting Infrastructures

EDITED BY

Deborah Cowen,
Alexis Mitchell,
Emily Paradis,
and Brett Story

UBCPress · Vancouver · Toronto

29 28 27 26 25 24 23 22 21 20   5 4 3 2 1

Printed in Canada on FSC-certified ancient-forest-free paper (100% post-consumer recycled) that is processed chlorine- and acid-free.

Library and Archives Canada Cataloguing in Publication

Title: Digital lives in the global city : contesting infrastructures / edited by Deborah Cowen, Alexis Mitchell, Emily Paradis, and Brett Story.
Names: Cowen, Deborah, editor. | Mitchell, Alexis, 1983- editor. | Paradis, Emily, 1968- editor. | Story, Brett, editor.
Description: Includes bibliographical references and index.
Identifiers:
   Canadiana (print) 2020025815X |
   Canadiana (ebook) 20200260480 |
   ISBN 9780774862387 (softcover) |
   ISBN 9780774862394 (PDF) |
   ISBN 9780774862400 (EPUB) |
   ISBN 9780774862417 (Kindle)
Subjects:
   LCSH: Cities and towns—Technological innovations. | LCSH: Technology—Social aspects. | LCSH: Sociology, Urban. | LCSH: City and town life. | LCSH: Smart cities. | LCSH: Online social networks—Social aspects. | LCSH: Information society.
Classification: LCC HT153 .D54 2020 |
   DDC 307.76—dc23

Canadä

UBC Press gratefully acknowledges the financial support for our publishing program of the Government of Canada (through the Canada Book Fund), the Canada Council for the Arts, and the British Columbia Arts Council.

This book has been published with the help of a grant from the Canadian Federation for the Humanities and Social Sciences, through the Awards to Scholarly Publications Program, using funds provided by the Social Sciences and Humanities Research Council of Canada. We also acknowledge support from Furthermore, a program of the J.M. Kaplan Fund.

Furthermore:
a program of the J.M. Kaplan Fund

Printed and bound in Canada by Friesens
Set in Gilam and Sabon by Artegraphica Design Co.
Copy editor: Lesley Erickson
Proofreader: Caitlin Gordon-Walker
Indexer: Judy Dunlop
Cover designer: Martyn Schmoll
Illustrations on pages 18–19, 114–15, and 206–7: Lize Mogel

UBC Press
The University of British Columbia
2029 West Mall
Vancouver, BC V6T 1Z2
www.ubcpress.ca

This book, and the collaborative research out of which it emerges, is anchored in the work of activists organizing for urban justice around the world. The collection took nine years to bring into being and materializes in 2020 to a world on fire, with longstanding urban struggles for Black lives, migrant rights, and racial, economic, environmental, and infrastructural justice reaching a boiling point. In the context of the global pandemic, political life and digital life are ever more and inextricably entangled. This collection archives this extraordinary moment and reminds us that activists, artists, and scholars were taking up these issues long before they were "laid bare" by COVID-19. We dedicate this work to the movements and communities organizing for change – those whose labour has gotten us to this definitive moment, and those who are working to take global urban life into a more just future.

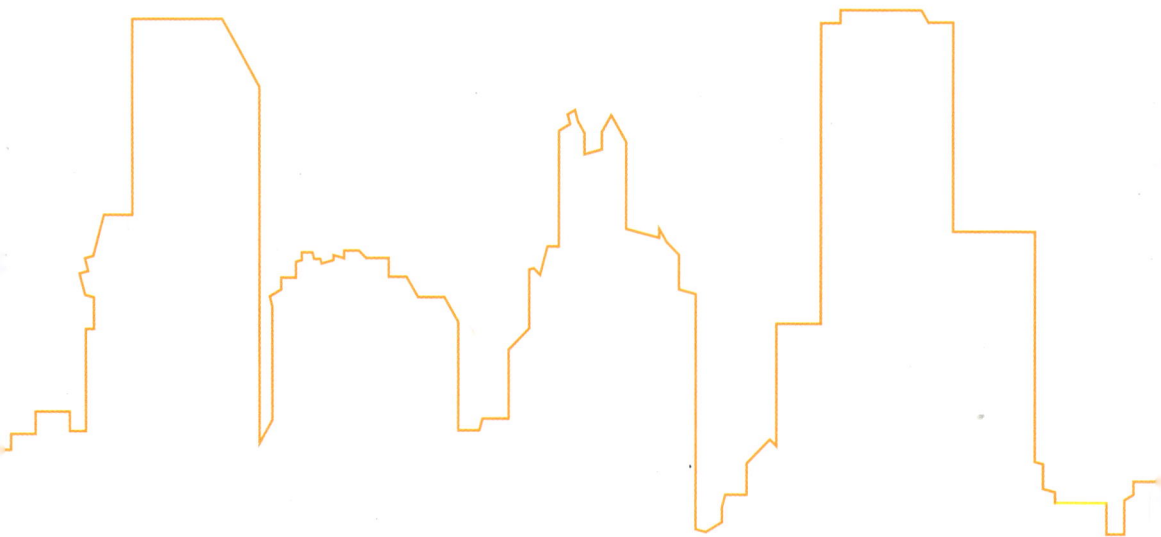

Contents

TORONTO

SECURITY AND SURVEILLANCE

The Towers in the World, the World in the Towers

THIS BOOK EMERGED out of a unique creative documentary project called *Highrise*, at the National Film Board of Canada (NFB). *Highrise* was a seven-year experiment in documentary, community-engaged research and cocreation in nonfiction storytelling using emergent technologies. The documentary makers who were part of the project were interested in the actual people who make up the growing density of the global suburbs (inspired originally by our own city of Toronto, Canada). We also wanted to learn how vertical lives – literally, residents of these suburbs living in high-rise buildings – are entangled with digital infrastructures and systems.

For *Highrise*, we did not follow the conventional documentary process of first finding subjects and experts to interview for a film, then interviewing them based on our questions, and then disappearing into an edit suite to shape an argument. We were mandated instead by the NFB to experiment in both form and content. Inspired by the NFB's legendary Challenge for Change project in the 1960s and 1970s, we were challenged to build the project out of a process rather than defining the process by the end goal. Our process was informed by community-based and cross-disciplinary methods of cocreation. The project grew out of relationships and community engagements rather than arriving with preset agendas.

For seven years, our team of documentarians worked alongside architects, urban planners, housing activists, technologists, scholars, and, most importantly, high-rise residents themselves. Together, we built relationships over time, and in so doing, we also jointly built the framing, the questions, and the goals of each of the many projects that spilled out from the work. Beyond documentary films, we created websites, installations, community newsletters, participatory media workshops, live performances, and place-based interventions.

We collaborated with the *New York Times*, Wired.com, the City Lab at *The Atlantic*, the *Globe and Mail*, the Mozilla Foundation, and others. The work was seen by millions and garnered two Emmys, a Peabody, and many other awards. Most important for us, the work was a living, digital example of how media makers can work more ethically and respectfully in cocreating work within communities and across disciplines to challenge the often exploitative and extractive modes of cultural production.

This book is most closely tied to the final documentary production in the *Highrise* collection – *Universe Within*. We had already been working for several years at the Toronto site of two high-rise buildings in a suburb called Etobicoke. To learn more about daily life in these high-rises from the perspectives of residents, the NFB supported a multiyear series of weekly participatory media workshops, and partnered with a pilot civic engagement project in the building by the United Way. The aim of the workshop process was to collaboratively design research and media projects that both responded to and reflected the needs and priorities identified by the people who lived in the high-rises. Our core group included six residents, three media makers, and the community engagement team (also residents) hired through the United Way office.

The team was remarkably consistent over the years, but a few key factors mitigated reaching full capacity for democratic process in the project. The funding that flowed was inequitable and precarious. While the project ultimately ran for many years, the funding was frequently up for renewal both at the United Way and at the NFB, creating a sense of instability and uncertainty. Further, the media makers and community-development officers were paid for their time, but the participants received only honoraria for their labour, which also amplified the inequities of race, class, and power in the group.

Finally, key decisions about programming, design, and production were pitched and made off-site at the institutions by nonresidents, which also limited the democratic capacity of governance in the group. To begin to address these constraints, our documentary team framed our request for NFB institutional support as support for a "process" rather than for a given media outcome, as most documentary projects work. In this way, the core decision making about editorial and distribution decisions and media outcomes were centred on and made within the group meetings and determined by consensus.

The group met weekly in one of the building's main floor community rooms, and many outcomes emerged out of the process we set up, including a community newsletter, public presentations, local events, and a series of wide-reaching NFB media projects. The first of these was a photo-blog called *The Thousandth Tower*. We also adapted the photo-blog into live spoken-word performances. The residents presented their photos and stories at the building to their neighbours, then at a large event sponsored by city hall downtown (an event attended by 350 people, mostly civil servants, politicians, designers, urban planners, and educators), and finally at a reading at Toronto's Harbourfront Centre.

The NFB also created a digital multimedia version of the project. From the presentations, a new collaboration emerged, this time with architects who joined the workshops for several months to work with residents to reimagine the outdoor spaces around the buildings. The resulting project, called *Millionth Tower*, was an interactive, animated documentary featuring stories, drawings, and ideas to reimagine the spaces around buildings. The neglected tennis court was reimagined as a play area, community gardens sprouted at the back of the buildings, and a market filled the parking lot. The group presented the documentary at community events throughout Toronto and to the world online, through publications with Wired.com and the Mozilla Foundation.

Although our core team spent a lot of time together, we had little connection to the two thousand plus residents in the buildings. Residents seemed to hurry to their apartments, rarely speaking with one another. As a team, we wondered about the digital lives and connectivity of the residents, and we decided to survey the building in a more systematic way. Together with scholars, we designed a participatory methodology, and we recruited a team of fourteen residents to conduct a peer-to-peer, door-to-door survey of their neighbours throughout the building. Collectively, the researchers spoke fourteen of the languages represented in the building, helping us to reach many residents who would not – or could not – speak with us.

It was illuminating to be part of this process – to see how a survey, when conducted by residents, could help neighbours get to know one another and begin working together to make their home a better place. The data gathered was also fascinating. Ninety-three percent of those interviewed had not been born in Canada. More than 50 percent were under the age of twenty. Eighty percent of households were connected to the internet, despite the financial burden. No one surveyed had a significant connection to someone in downtown Toronto – only twenty kilometres away – but many had daily contact with friends and family in other suburbs, and all were linked to loved ones on the other side of the world.

The peer-to-peer process of more equitable data collection also proved empowering to residents. Many surveys take months – even years – to process data and share results. But we made it a priority to share the data with the residents within a few weeks and to involve them in its analysis through multiple roundtables. They immediately used the information to successfully advocate for a much-needed new playground for their children.

That early fruitful research formed the basis for a much broader academic and documentary collaboration called Digital Citizenship in the Global Suburbs, which later became *Universe Within*. Deborah Cowen and Emily Paradis of the University of Toronto secured a grant from the Social Sciences and Humanities Research Council (SSHRC) to develop our partnership. We became an interdisciplinary team of academic researchers, documentarians, and peer researchers.

We expanded our scope. We surveyed the scholarly literature and news media, we interviewed key informants, and building on the networks we had established over seven years of sustained work on *Highrise*, we found compelling and complicated sites where the vertical intersects with the digital in Africa, South America, Asia, Europe, and North America.

In Baku, Azerbaijan, for example, we developed relationships with bloggers and peace activists, who then documented their own work for our documentary to show how they are using online tools to create safe virtual spaces for opening dialogues between youth in Azerbaijan and Armenia.

In San Cristobal, Venezuela, we worked with local social activists to document one of the early uses of WhatsApp, a closed-group mobile application that was used to securely connect residents of a building when civil unrest erupted in the neighbourhood and their building became the target of pro-government militia groups.

In Guangzhou, China's third-largest city, we worked with LGBTQ+ activists to safely document how a woman named Ling had used social-media sites to arrange a "mock marriage" between herself and a gay man, all to conceal her lesbian identity from her parents, with whom she lived in a high-rise apartment.

These documentaries were produced remotely. Local photographers, storytellers, and activists worked on the ground within communities and corresponded over the web and via phone with our editorial team in Toronto. Relationships were based on trust and vision developed over many years of network building as well as the track record of the earlier *Highrise* projects. Editorial decisions were situated within critical and ethical concerns, respecting the safety of the storytellers and story holders, prioritizing self-representation, and centring the perspective of those who lived directly within the communities.

In tandem, the academic team chose three major sites – Toronto, Mumbai, and Singapore – for in-depth research, the outcomes of which are presented in this book.

The resulting online documentary, *Universe Within: Digital Lives in the Global Highrise*, was built from technologies that were, in 2015, new and experimental but that now dominate in the field of emergent storytelling: WebGL, Depthkit, and Kinect point-cloud data capturing. Most importantly, the stories told in the piece were prescient. From Toronto to Singapore to Mumbai and many other cities, we learned that, for our species, both the digital and vertical are becoming inescapable. As we race towards more digital integration with artificial intelligence, virtual reality, surveillance, big data, and robots, we are also experiencing the rampant vertical development of our cities, a process that is shuffling, condensing, and displacing millions of people. Whom do these processes exclude? Who wins? Who loses? And how might we harness these new technologies to improve our collective future?

The *Highrise* experiment now animates many of the questions of my current work at the Co-Creation Studio at the MIT Open Documentary Lab. We have completed a major field study, titled "Collective Wisdom," for which we spoke with 160 practitioners about their own cocreative practices in documentary, art, journalism, and adjacent disciplines. Co-creation methodologies can help link media-making to justice and equity and contest the often extractive processes of conventional documentary making and academic research. We arrived at the following definition of cocreation. Cocreation is a constellation of methods that function as alternatives to singular authorship and singular authority; it involves projects that originate from more than one person, discipline, or system. Cocreative media practices are underfunded and underrecognized, but they allow for new, better questions, ones for which there are not always single answers. It also emerged that cocreation carries with it significant risks: it can compromise editorial integrity, it can set up expectations for outcomes that are not attainable, and it can lead to the exploitation of labour, the theft of ideas, and the perpetuation of inequities that the cocreators had claimed to address.

The notion of cocreation can also be co-opted by bad actors, especially within technology spaces. But practised within ethical frameworks, and by explicitly addressing these risks, cocreation can enrich daily practice: it demands self-reflection and can forge more harmonious, transparent equitable relationships between partners, within and across communities, beyond disciplines, and working with nonhuman systems, many of which we do not yet fully understand.

KATERINA CIZEK, *director of the NFB's* Highrise *project*

When Localities Go Global

MUCH OF THE CONCEPTUALIZING of the local in both general and academic writing has been centred on physical or geographic proximity. This marks the local as a sharply defined territorial zone, with the associated implication of particularity and closure.

This collection and the project upon which it is based looks at the world from the perspective of the suburban high-rise. And it is this unusual focus that makes its contribution genuinely useful and exciting. It has the effect of exploding the boundedness of the "suburb." The suburb becomes the "global suburb," and its residents turn out to be highly mobile even when sitting on their suburban lawns. A similar analysis can be made for localities within cities, notably neighbourhoods. This intersection of digital interactive space and the deeply localized transforms both "the digital" and "the local" into unstable categories. Neither term fully captures the condition it seeks to name. There was a time when these terms had fixed meanings. Now we know that such precision is merely rigidity.

This instability of meaning unsettles the notion of the suburb. The suburb becomes a variable. It can be the traditional suburb, and in much of the world it remains that way. Or, at the other extreme, the suburb becomes a novel type of space, one constituted both as marked terrain and as a borderless technical zone.

It is this possibility of a borderless technical zone that demands interrogating. One way of framing this interrogation is to ask whether locality is a condition that is actually lost or erased when we speak of the global suburb. Does the global suburb disintegrate into a digital free-for-all? Or is it a specific type of technical space, one that incorporates and is partly marked by thick local contents? In short, could it be a counterintuitive type of space, one that includes both the generic technical condition and a bounded sort of communication dynamic?

In my own research, I have found over and over again that interactive digital space cannot be reduced or flattened to the technical capacities at work. In other words, emphasizing only the technical features being mobilized will tend to transform the digital, interactive space into a generic technical zone. To interrogate this narrow technical understanding, I found it helpful to compare similar types of interactive digital spaces used by diverse types of actors – as diverse as financial traders, on the one hand, and civil society organizations, specifically, human-rights activists and environmental activists, on the other. Even as these two types of actors used similar technical capacities, there was a larger ecology of meaning that got mobilized and enacted in that technical space, that emerged partly from the particularities and specific cultures of the localities of participants. They marked the difference. The localities in this case were, on the one hand, the places where these local human rights or environmental activists were pursuing their struggles and, on the other hand, the specific financial markets within which the traders worked.

A purely technical analysis of the capabilities mobilized for those interactions cannot pick up on this type of specificity. And in not picking up on this specificity and simply focusing on the technical capacities, the observer is kept from recognizing the role of locality even in global digital networks. Human-rights activists around the world do their work in enormously diverse national settings, and within each one we will find multiple diversities as well. Similarly, financial traders in gold as a commodity operate in a different space than financial traders in aluminum. Capturing such specificities is a whole task unto itself, easily neutralized by an exclusive focus on the technical functions getting mobilized.

But to get back to the global suburb – does the presence of diverse local ecologies of meaning also matter in understanding this emergent condition?

When multiple territorial localities – city neighbourhoods and suburbs – become part of a digital interactive space, the question of membership can take on whole new meanings. If we consider, for instance, the possibility that certain causes are shared at least by some people in cities and suburbs throughout much of the world (examples might be environmental protection, better public transport, or freedom of speech), then a new kind of commons can get generated. And the participants of this commons may eventually constitute a kind of informal citizenship through their online and in situ work around such causes.

What gets constituted through this work is a horizontal digital space marked by the recurrence of particular features: multiple localities integrated via recurrence rather than hierarchical organization. Further, because the space remains local, it is local issues that constitute that global space, rather than national, international, or global issues. The local gets a chance to become constitutive of meaning in a global space. This takes on additional meaning against the background of disassembling nation-states, a subject I elaborate on at length elsewhere. In this context, urban spaces of diverse kinds – from cities and their neighbourhoods to suburbs and even functioning regions – emerge as strategic sites for rebuilding specific aspects of citizenship.

These are some of the issues and questions explored in this collection, which makes an exciting contribution to this and a wide range of linked issues. It does so on its own terms, with its own concepts and aspirations. This is simply my bridge into this world where the territorial and the digital intersect.

SASKIA SASSEN, *Robert S. Lynd Professor of Sociology at Columbia University*

Digital Lives in the Global City

Introduction

Deborah Cowen, Alexis Mitchell, Emily Paradis, Brett Story

OUR TEAM OF RESEARCHERS and filmmakers stands between two tall buildings. These buildings, once a model of the modernist "towers in the park" style, now inhabit a poorly lit space criss-crossed with fences and crumbling sidewalks.

The surrounding neighbourhood also shows signs of disinvestment: retail space is sparse, public transit is inadequate, and decent employment opportunities are almost nonexistent.

Inside the towers, infrastructure fails: elevators malfunction, heating systems struggle, walls grow mould.

But these towers are also home to hundreds of remarkable residents from all corners of the earth. They pay steep rents to private property owners. The vast majority were born outside of the country. More than half have migrated to Toronto within the last decade. Collectively, they speak dozens of languages. Here, survival can depend on digital infrastructures, on the ability to use them to stay connected to the world and, when necessary, to resist the growing inequalities of power and wealth these technologies enable.

Welcome to life in the global suburb.

In 2008, residents of two buildings in the suburban neighbourhood of Rexdale, Toronto, became partners in the National Film Board of Canada's *Highrise* project, a series of interactive web documentaries that charted how the global explosion of high-rise living was changing the urban experience. Over the next seven years, working in collaboration, teams of residents, documentary filmmakers, and scholars, including ourselves, explored the realities and possibilities of life in these towers through photography, design, video, and the creation of interactive online environments. This book is a unique outgrowth of that project, particularly the final online documentary, *Universe Within: Digital Lives in the Global Highrise.*

While working on *Highrise*, we learned that the two buildings had much in common with high-rise housing throughout Toronto's inner suburbs. In contrast to the increasingly white, professional downtown core, the suburbs have become diasporic zones where poverty goes hand in hand with newcomer or racialized status. Decades of inner-city gentrification and rising housing prices have pushed communities of colour to inner-suburban areas such as Rexdale. The suburbanization of lower-income, racialized, and otherwise marginalized communities is a phenomenon shared by many cities in North America and around the world. In a sense, it is the flip side of gentrification, and it is deeply bound up with financialization (the growth and dominance of financial institutions and markets) and the globalization of urban real estate markets.

This project was born in a very particular time and space; 2011 was a year of extraordinary political revolt around the world. Mainstream and social media featured a steady stream of accounts, images, and analyses of protest movements erupting across the globe. In the *Guardian*, for instance, John Harris reflected on this conjuncture of popular uprisings, and the Occupy movement in particular, and identified a series of other place-specific revolts or "convulsive events" in "Tunisia, Egypt, Libya, Bahrain, Syria, and Yemen. Mass protests against economic breakdown and austerity in Greece, Italy, and Spain. Marches and protest camps in Chile and Israel."[1] Revolts marked 2011 as exceptional not only because of their quantity but also because of their particular qualities. Writing in the *Financial Times* on August 29, in an article titled "The Year of Global Indignation," Gideon Rachman emphasized common provocations: "Many of the revolts of 2011 pit an internationally connected elite against ordinary citizens who feel excluded from the benefits of economic growth, and angered by corruption."

These claims about the deepening nature of global economic divides are well supported by research. In advance of the 2015 World Economic Forum, Oxfam launched a major report on global inequality that highlighted that the world's eighty-five richest people are now as wealthy as the poorest half of the global population. Directly echoing the Occupy movement, Oxfam elaborated that "the combined wealth of the richest

1 percent will overtake that of the 99 percent next year unless the current tide of rising inequality is checked."

Although deepening economic inequality was a pervasive theme across many of these revolts, their specific contexts also mattered: some groups were resisting authoritarian states, others were protesting colonial occupation, and still others were critiquing the dramatic polarization of wealth and the privatization of public spaces, services, and infrastructures that define contemporary urban life in so many contexts. These protests were not the same, but they were connected: they were all responses to different incarnations of globalized capitalism, imperialism, and authoritarianism. They were all connected through what geographers might call geopolitical economic violence.

The 2011 revolts were also connected literally and materially through digital technologies. In his article, Rachman commented on the "mysterious" nature of a "global mood" that the revolts were both reflecting and creating. It is true that the complex and multiple sources of any mood will remain a mystery. But we recognized that the creation of a global mood on this vast scale depended on real links and associations – in other words, on infrastructures: the material and social systems that make daily and human life possible or, as Brian Larkin famously wrote, the "matter that enable the movement of other matter."[2] As infrastructures have multiplied, expanded, and intensified over time, they have become critical to modern life, and by "critical" we mean as necessary as air or water.[3] It is almost impossible to imagine life today without the infrastructures that connect us. When infrastructure works, it often goes unnoticed and is devalorized because, as Bruce Robbins has argued, it both belongs to the public domain and "smells" of the public. When it fails, it becomes visible, and we become aware of our dependence on it and on just how interlinked we have become.[4]

But even as infrastructure and technologies connect us, they also divide us because access to them is deeply uneven, creating hierarchies of digital citizenship based on class, race, education, gender, age, and location.[5] North Americans have the highest rate of access while Africans have the lowest, yet there is dramatic unevenness within regions and cities.[6] For instance, in American cities, telecommunications companies often engage in "electronic redlining," actively limiting the infrastructure available to "minority neighbourhoods."[7] Uneven access to infrastructure creates what Doreen Massey in her classic 1991 essay "A Global Sense of

Place" referred to as "power geometries," stark divisions between rich and poor regions and between social classes.[8] Digital technologies have helped shift economic and political power to the private sphere at the expense of the state, women, workers, and citizens in general, making private market ownership and operation of infrastructures a defining feature of the government of digital citizenship, particularly in the global south.[9]

The revolts of 2011 demonstrated that digital technologies, as both a tool for protest and a source of division, were deeply woven into contemporary political and urban life. Commentators pointed to the instant power of "BBM," or Blackberry Messenger, to draw people together in the streets during the London riots, and they highlighted Twitter's and Facebook's role in sustaining the transnational networks that underpinned the Arab Spring in the cities of Egypt, Tunisia, and beyond.[10] In these cases and others, the power of the digital to alter the landscape of political acts and identities was brought into focus. So-called internet activism was lauded for its capacity to transcend the traditional limits of state sovereignty and space: activism could now take place in public or private spaces, from the centre of Tahrir Square or from a living room halfway around the world.

Digital technologies were fostering connections not only within particular geographic contexts and movements but also between them. Social media in particular seemed to be allowing groups to connect across different social and spatial locations and, in doing so, to transform the relationship between people, places, and politics. Indeed, the movements of 2011 were connected in at least one more profound way – through geographies that were overwhelmingly urban and simultaneously transnational. Acts of revolt were defined by an explicitly urban geography and lexicon – that of Cairo, Tunis, Oakland, London, Hong Kong, or Athens. Urban sites of symbolic, political, and economic importance – sites such as squares, bridges, roads, and parks – were stages for many of the most dramatic standoffs between protesters and police. And coalitions across movements highlighted the transnational nature of this urban geography – for instance, through Twitter and Facebook, acts of solidarity could circulate between New York and Cairo or between Oakland and Gaza. The protests made it clear that digital infrastructures are not simply located within urban centres – they literally constitute contemporary urban life. The physical infrastructures of the city offer the spaces and material structures of digital connectivity, and digital technologies, in turn, provide the infrastructures of circulation and communication for the city. When people take control of digital technologies in these spaces, they, too, become infrastructure in Larkin's sense of "matter that enables the movement of other matter."[11]

Today, cities are magnets for migrants, spaces of contestation, targets of state and nonstate terror, objects of securitization, built forms for speculation and surplus capital, and media for the mass accumulation and mass dispossession of wealth. Likewise, urbanization has become such a definitive feature of modern life that disciples of the so-called urban revolution believe that it now drives global accumulation, stealing the lead from industrialization. Scholars are now engaged in lively debate about planetary urbanization, a thesis that suggests that cities are not simply home to most of the planet's human population but, more importantly, set the tone and pace for global social, political, and economic life.[12] Consider that you don't even need to live in a city to feel the effects of planetary urbanization. Nor do you need to live with apps and devices to feel the effects of digital life. The digital and urban revolutions mean that events that happen in one part of the world are profoundly and instantly connected to other places.

· · · · ·

To understand how these trends were playing out in the lives of suburban residents in Toronto, in 2011 we began working with fifteen residents of the two high-rise buildings. Over six weeks and using fourteen different languages, resident-researchers went door to door and, with their neighbours' consent, undertook a substantial survey of their digital lives. We then repeated the survey in 2014. We learned that although the residents' lives were financially precarious, almost all households had internet access in their homes: in 2011, the rate was 80 percent, and by 2014 it had increased to over 90 percent.

High-speed internet is notoriously expensive in Canada, yet residents were making it a priority in their household budgets. A substantial portion, as many as 25 percent, were sacrificing food to afford it. Debt was also clearly becoming a key means through which members of these communities were providing this critical but privatized infrastructure for their families and loved ones. We visited a young refugee who had no furniture or decorations in her apartment except for a computer, the desk upon which it sat, and a small bed in the corner. All of her resources went towards her digital expenses first. In fact, residents likened the internet to water or air. We learned that residents used the internet for profoundly "extended" intimacies – to virtually sleep with their loved ones, meet their grandchildren for the first time, or spend a day with family members back home.

What we discovered flew in the face of much of the standard research on digital divides, which posits that low income and low access are inevitably paired. The residents' stories demonstrated that their local and transnational lives were profoundly supported by digital technologies. People's lives and loved ones are now stretched across the globe and materially underpinned by email and Skype and by electronic money transfers and digital news media. On the global scale, diaspora demands the digital.[13] Likewise, at the local level, in the city – where connecting with members of your community, getting to a job, or running errands can be deeply challenging and time-consuming, in no small part due to a lack of public space and poor transit options – digital technologies and connectivity are essential.

If the explosion of revolts in 2011 pushed us to think more carefully about the implications of urban and digital transformations for modern life and citizenship, this less sensational finding caused us to see that focusing on spectacular events can overshadow sustained investigation of the more subtle and quotidian ways that the digital is reshaping urban life.[14] In all the speculation, analysis, and celebration that followed the events of 2011, many important questions remained unanswered. At the level of everyday life, is being political now also a question of being digital?[15] More crucially, even though we can't see it, how is networked urbanism sculpting our political landscapes?

Our observations of digital life in the two suburban high-rises in Toronto prompted us to expand our focus. We initiated research and documentary work to explore (1) how digital technologies are remaking urban space and urban life in other places and (2) how people are making use of those same technologies to organize alternative futures. This collection draws on the extensive fieldwork and multiple research partnerships we entered into in Toronto and two other important global cities – Mumbai and Singapore – but it also mobilizes more targeted documentary work completed in a dozen other cities.

Viewing digital technologies and urban environments as infrastructures that can both connect and contain us, we set out to answer broad questions about political life today. How are transformations in digital technologies and urban space remaking citizenship and political life? How does thinking about the digital and the urban as critical and intersecting infrastructures facilitate creative thought about agency, identity, and subjectivity in this moment and about uneven global circulations of power and capital? Indeed, because access to infrastructure is organized along the lines of class, race, gender, education, age, and location, some people and groups have no direct access to the infrastructures of connection, and yet they are still governed by them, an inequality that influences the stories we tell, how we tell those stories, and who is able to tell them in the first place. These are the classic problems of positionality and reflexivity that feminist scholars have long grappled with, and we return to these questions after first outlining our investment in this work.

To ensure that we would foreground the most pressing stories, we adopted Richa Nagar and Susan Geiger's approach to fieldwork and asked: How can we "produce knowledges across multiple divides without reinscribing the interests of the privileged?"[16] Like Nagar and Geiger, we strove to situate knowledges and solidarities *in place*. As they argue, where we live determines how we think about the world, whom we ally with and feel committed to, and how we participate in movements for social change. Our fieldwork sought to cultivate situated solidarities, not only between the research team and local communities but also among local actors.

In Toronto, Singapore, and Mumbai, we collaborated with local scholars, activists, artists, journalists, and grassroots organizations to plan and execute the research and documentary work. We wanted to avoid an extractive model of research and instead spark conversations, networks, and products (such as this book) that would advance local projects of resistance and endure beyond the project. Over Skype and by email, we worked with partners in Mumbai and Singapore for months before our fieldwork began, planning a research and documentary process that would have intrinsic benefits for everyone involved. With their input, we selected research strategies that met local needs, from traditional interviews and focus groups to arts-based methods such as Photovoice. Once our team was on the ground in each city, we worked closely with partners to respond to emergent stories and actions, modifying plans as necessary and making research materials and processes available for use by local actors.

In Mumbai, for example, we based our inquiries in three different places. One was an affluent building in South Mumbai that came to our attention via its residents' sophisticated and well-resourced social-media campaign. The other two were working-class neighbourhoods in Mira Road, a distant commuter suburb whose residents had considerably less access to the tools, languages, and infrastructures of advocacy. At the project's closing forum, we brought together residents from all three sites, along with local scholars and nongovernmental organizations, to share information about their local struggles and to consider collective responses. The connections made that day reverberated in ways we could not have predicted. More than two years later, one of the planning scholars we had worked with was invited to present a social-impact statement on a proposed rezoning in Mira Road. Her intervention drew on the experience of residents at one of the sites.

This volume features the work of celebrated scholars, but we also include a range of voices, including those of the research participants. The diversity of the contributors reflects, we believe, our commitment to understanding both global transformations and the experiences of those on the ground, including, centrally, the perspectives of people who are actively navigating, and often shaping these transformations. Text is one key element of the book, but much like the experience of modern, digitized urban life, it is not the only narrative medium. We also mobilize some of the best elements of the online experience. Images are not only plentiful and a product of our partnership with the National Film Board of Canada but also a means through which connections are drawn, arguments are elaborated, and stories are told. Maps and other forms of data visualization such as infographics are also interspersed through the essays, which are visually appealing and of varied lengths.

We start in Toronto, in the Rexdale high-rises where our project began. In "Digital Debt in a Precarious City," scholar Emily Paradis and filmmaker Heather Frise map the digital divide onto the city. They show that, on the one hand, residents are grappling with the digital as a critical infrastructure of everyday life, a medium for sustaining intimate ties, and a tool for the reinvention of transnational citizenship. On the other hand, digital technologies are also a mechanism for the extraction of wealth from – and the transfer of risk and debt onto – low-income households and neighbourhoods. The pieces that follow zoom in. Geographer Alan Walks surveys the city's debtscape, mapping the uneven social and spatial distribution of debt and its link to financialization. Sociologist Krystle Maki investigates the digitally enabled surveillance of Ontario welfare recipients, showing how workfare regimes are applying new technologies to the age-old functions of gendered and racialized moral regulation of the urban poor. Organizer Judy Duncan takes us inside ACORN Canada's campaign for affordable internet access led by low- and moderate-income community members. Finally, in "Transmutations," scholar and writer Nehal El-Hadi offers a textual assemblage – a collage of voices and ideas – through which she pieces together a theory of the production of presence and reveals how women of colour activists in Toronto are using online activities to claim their right to the city.

Rather than proceeding straight to Mumbai, we take a detour to explore the twinned themes of security and surveillance and what they reveal about digital technology and urban transformations. In "Digital Medieval," geographer Stephen Graham takes a global view of military tactics, positing that new digital surveillance and tracking technologies are producing fortified, urban enclaves and camp-like enclosures in and between cities around the world. Author and educator Simone Browne and securitization scholar Joshua Scannell narrow the lens. Browne explores the connections between the invention of closed-circuit television equipment and the ubiquitous surveillance of African American urban neighbourhoods. Scannell examines the rise of megadata technologies and their use by US police forces, in particular the development of an NYPD initiative called the Domain Awareness System and its integration into so-called predictive policing programs. In "Policing Border through Sound," activist scholar Anja Kanngieser widens the focus once again to investigate the sonic governance of space. Writer and activist James Kilgore's piece on the rise of electronic monitoring by correctional authorities reinforces Scannell's criticisms of digital technologies as new tools of racialized urban social control. Finally, Visualizing Palestine, a graphic research collective, offers a set of infographics that showcase the digital apartheid that characterizes and organizes the lives of Palestinians in the Occupied Territories.

We then travel to Mumbai to explore the financial and digital infrastructures that are organizing dispossession and resistance in the outer geographies of this globalizing city. In "Mumbai Rising, Buildings Following," scholars Emily Paradis, Brett Story, and Deborah Cowen employ three case studies to examine the intersection of finance capital, digital media, and protest as they play out in the lives, and buildings, of diverse Mumbai residents. Our understanding of the complexities of global real estate capital, digital infrastructure, and urban planning is deepened by architects Hussain Indorewala and Shweta Wagh's piece on Mumbai's recent development plans and the market-friendly planning regime they secure. To better understand a community's battle for their homes in the contested suburb of Mira Road, geographer Deborah Cowen interviews Kashaf Siddique, millennial Mumbai activist and resident. Filmmakers Paramita Nath and Deborah Cowen report from the frontlines of a heated standoff between community members and the local state when, following an expansive social media campaign, residents barricaded themselves inside their Mumbai highrise compound to protect their building from demolition. Finally, sociologist Shilpa Phadke and independant journalist Sameera Khan's contribution connects this section back to the city and digital technologies.

Before proceeding to Singapore, we stop to reflect on the many ways that technology is scripting and shifting how people are experiencing the urban environment, how it is changing not only our material worlds but also our emotional and experiential worlds. The section opens with "High-Altitude Protests and Necropolitical Digits," in which geographer Ju Hui Judy Han maps activism and protests in Korea, and the section ends with communications scholar Nicole Starosielski's "Network Dislocations," which, by tracking the movement of submarine cables towards and away from urban hubs, makes visible the often invisible technologies that structure our daily lives. In between, artist Indu Vashist shows us how taxi apps are shaping everyday movements and politics in India; media studies scholar Shaka McGlotten delves into the complexities at the intersections of queer performance and digital life in "The Most Hated Woman in Israel," a reference to YouTube performance artist Natali Cohen Vaxberg; and artist and researcher Heather Frise explores the use of alternative internet networks to overcome the problems of network affordability and corporate and government control.

Finally, we arrive at Singapore, a city known for its billion-dollar high-rises, shopping malls, and technological advancements. Scholars Deborah Cowen and Alexis Mitchell dig past the veneer of a city-state built on progressive housing mandates to look at the often forgotten and exploited foreign labourers who work to build and staff the city's high-rises and whose experiences speak to the complex ways technological advancements are both structuring and hindering their everyday lives. In "Skyline of Dreams," photographer Grace Baey relates the heavy costs and physical burdens of the job, including injuries, negligible support, and being forced into precarious positions. Political scientist Charmaine Chua's "Sunny Island Set in the Sea" shifts the focus to the base material of the cityscape – sand. By examining its geopolitical prominence in both the construction and technological sectors, Chua homes in on the exploitation, global inequality, and environmental destruction associated with this ubiquitous material. Geographer Natalie Oswin switches the focus from material narratives of the city to the social and political, writing about the city's grand narratives and checking them against the realities of foreign domestic labour, sexual health practices, and heteronormative city regulations. Lastly, geographer Symon James-Wilson showcases the creative ways foreign labourers are relying on digital technologies to undo or rescript dominant narratives. Each piece in the section opens a new window and perspective on a city that prides itself on its stories. Each is a creative intervention to renarrate the everyday experience of living and working in Singapore.

· · · · ·

At first glance, the issues and conflicts we document in these pages appear to be quite diverse. Digital debt in Toronto. Struggles over unsafe and illegal buildings in Mumbai. The conditions of migrant work in Singapore. They appear to be disparate, yet they are symptoms, or diagnostic events, of globalized digital life. The digital has clearly emerged as a critical infrastructure in the modern era. Indeed, digital technologies are opening up myriad mundane ways for people to temporarily fix a whole range of pervasive (perhaps systemic) contemporary problems, including the dislocations and separations caused by the urban revolution, the movement of people around the world, and the deeply classed and segregated nature of urban space. In this context, digital technologies are having profound yet contradictory impacts. On the one hand, they are creating new challenges and deep injustice. On the other, they are allowing new forms of human creativity and connectivity to flourish.

Without a doubt, vastly different histories and future trajectories are defining cities around the world. Some cities are shrinking while others are experiencing dramatic rates of growth. Some are experiencing massive deindustrialization while others are seeing extraordinary industrial expansion. In some regions of the world, informal settlement is the dominant style of urbanization; elsewhere, centralized planning remains the norm.[17] Nevertheless, despite extraordinary variation in the physical, social, economic, political, and cultural shape of cities, it is clear that urban growth and change are transnational in scale, and it is now difficult, if not impossible, to imagine contemporary urban life without the digital. Cities have become networked environments. Almost all cities in the global north or south contain the physical traces of digital connectivity. Their built form is increasingly designed to accommodate the infrastructures and cultures of digital life. Wi-Fi zones, cable boxes, charging stations, and surveillance systems are all fragments of these forms. In a polarized city such as Mumbai, billboards advertising mobile phones cover the flyovers that stitch together the urban fabric. In "slum" settlements of the city's urban core, satellite dishes carpet the rooftops of informal housing while high-rises fill the horizon on the urban fringe. In Singapore, migrant construction workers, who labour under atrocious conditions, use their mobile phones to send a dollar at a time home to their families in Bangladesh, when they can send anything at all. In Toronto, private digital networks have become critical infrastructure and drive up debt in already stressed diasporic suburban communities.

Yet, as this collection shows, the impact of digital technologies and infrastructures are profoundly uneven. Digital divides reflect social divisions happening in other realms, and digital technologies can amplify them or even create new forms of disparity. Digital urban life is being forged through contradictory forces and feelings. To put it simply, digital technologies divide and connect, harm and help, often at the same time. The digital cannot be simply embraced or entirely shunned.

The contradictory experience and effects of digital technologies surrounded us in our research. In Singapore, for instance, we met domestic workers from Indonesia who were aggressively monitored by their employers through information and communications technologies such as closed-circuit television and cellphones. With few other employment options that allow them to earn a living and sustain families back home, many of these women said they had little choice in the matter. But many also expressed deep attachment to these same digital devices. They highlighted how they could occasionally call their daughters and sisters on the same device used to monitor their movements, a contradiction that provoked conflicting feelings of attachment, desire, and alienation. The gendered intimacy of domestic labour relations and the blurred boundaries of public and private that domestic workers navigate, are not only extending to ICT, but being reconstituted by it. Thus, the contradictory nature of connectivity is not simply about some people having better access than others but about the intimate "internal" divides that technologies and infrastructures orchestrate. Embedded in technology is design, and design makes some things more possible than others. But design can be overwhelmed. People can repurpose. So we launch into this journey, resisting utopian and dystopian impulses while nevertheless holding a firm critical eye on the workings of digital sub/urban life.

NOTES

1 John Harris, "Global Protests: Is 2011 a Year That Will Change the World?," *The Guardian*, November 15, 2011.

2 Brian Larkin, "The Politics and Poetics of Infrastructure," *Annual Review of Anthropology* 42 (2013): 329.

3 Anne Spice, "Fighting Invasive Infrastructures: Indigenous Relations against Pipelines," *Environment and Society: Advances in Research* 9, 1 (2018): 40–45.

4 Susan Leigh Starr, "The Ethnography of Infrastructure," *American Behavioral Scientist* 43, 3 (1999): 377–91; and Bruce Robbins, "The Smell of Infrastructure: Notes toward an Archive," *Boundary* 234, 1 (2007): 25–33.

5 See Omar Salamanca, "Road 443: Cementing Dispossession, Normalizing Segregation and Disrupting Everyday Life in Palestine," in *Infrastructural Lives: Urban Infrastructure in Context*, ed. Stephen Graham and Colin McFarlane, 114–36 (New York: Routledge, 2015); Majed Akhter, "Infrastructure Nation: State Space, Hegemony, and Hydraulic Regionalism in Pakistan," *Antipode* 47, 4 (2015): 849–70; Vyjayanthi Rao, "Infra-City: Speculations on Flux and History in Infrastructure-Making," in *Infrastructural Lives: Urban Infrastructure in Context*, ed. Stephen Graham and Colin McFarlane, 39–58 (New York: Routledge, 2015). On the uneven distribution of infrastructures and technologies, see Tracy Kennedy, Barry Wellman, and Kristine Klement, "Gendering the Digital Divide," *IT and Society* 1, 5 (2003): 149–72; Hiroshi Ono and Madeline Zavodny, "Gender and the Internet," *Social Science Quarterly* 84, 1 (2003): 111–21; Jan A.G.M. Van Dijk, *The Network Society: Social Aspects of New Media* (London/New

Delhi: Thousand Oaks/Sage, 1999); Jan A.G.M. Van Dijk and Kenneth Hacker, "The Digital Divide as a Complex and Dynamic Social Phenomenon," *Information Society: An International Journal* 9, 4 (2003): 315–26; and Mark Warschauer, "Reconceptualizing the Digital Divide," *First Monday* 7, 7 (2002): https://firstmonday.org/article/view/967/888.

6   International Telecommunication Union, *Measuring the Information Society 2012* (Geneva: International Telecommunication Union, 2012), http://www.itu.int/en/ITU-D/Statistics/Documents/publications/mis2012/MIS2012_without_Annex_4.pdf; and Internet World Stats, "Internet World Stats: Usage and Population Statistics," http://www.Internetworldstats.com/stats1.htm.

7   Chad M. Khal, "Electronic Redlining: Racism on the Information Superhighway?," *Katharine Sharp Review* 4 (Winter 1997): 3–9.

8   Doreen Massey, "A Global Sense of Place," *Marxism Today* 38 (1991): 24–29.

9   Barbara Crow, "Digital Restructuring: Gender, Class and Citizenship in the Information Society in Canada," *Citizenship Studies* 4, 2 (2000): 207–30; and Engin Isin and Evelyn Ruppert, *Being Digital Citizens* (Lanham, MD: Rowman and Littlefield, 2015).

10 Prithi Yelaja, "U.K. Riots Reveal Social Media Double Standard," *CBC News*, March 28, 2012; Nezer AlSayyad and Muna Guvenc, "Virtual Uprisings: On the Interaction of New Social Media, Traditional Media Coverage and Urban Space during the 'Arab Spring,'" *Urban Studies* 52, 11 (2015) 2018–34; and Philip N. Howard, Aiden Duffy, Deen Freelon, M.M. Hussain, Will Mari, and Marwa Maziad, "Opening Closed Regimes: What Was the Role of Social Media during the Arab Spring?," 2011, SSRN, https://papers.ssrn.com/sol3/papers.cfm?abstract_id=2595096.

11 Martin Coward, "Network-Centric Violence, Critical Infrastructure and the Urbanization of Security," *Security Dialogue* 40, 4–5 (2009): 399–418; Colin McFarlane and Jonathan Rutherford, "Political Infrastructures: Governing and Experiencing the Fabric of the City," *International Journal of Urban and Regional Research* 32, 2 (2008): 363–74; Simone AbdouMaliq, "People as Infrastructure: Intersecting Fragments in Johannesburg," *Public Culture* 16, 3 (2004): 407–29; and Julia Elyachar, "Next Practices: Knowledge, Infrastructure, and Public Goods at the Bottom of the Pyramid," *Public Culture* 24, 1 (2014): 109–29.

12 Henri Lefebvre, *The Urban Revolution* (Minneapolis: University of Minnesota Press, 2003); Neil Brenner and Christian Schmid, "Planetary Urbanization," in *Urban Constellations*, ed. Matthew Gandy, 10–13 (Berlin: Jovis, 2012); Michelle Buckley and Kendra Strauss, "With, Against and Beyond Lefebvre: Planetary Urbanization and Epistemic Plurality," *Environment and Planning D: Society and Space* 34, 4 (2016): 617–36; R.N. Reddy, "The Urban under Erasure: Towards a Postcolonial Critique of Planetary Urbanization," *Environment and Planning D: Society and Space* 36, 3 (2018):

529–39; and Ananya Roy, Eric Sheppard, Vinay Gidwani, Michael Goldman, Helga Leitner, and Anant Maringanti, "Urban Revolutions in the Age of Global Urbanism," *Urban Studies* 52, 11 (2015): 1947–61.

13 Nadia Caidi, Danielle Allard, and Lisa Quirke, "The Information Practices of Immigrants," *Annual Review of Information Science and Technology* 44, 1 (2010): 491–531; and Koen Leurs and Sandra Ponzanesi, "Communicative Spaces of Their Own: Migrant Girls Performing Selves Using Instant Messaging Software," *Feminist Review* 99, 1 (2011): 55–78.

14 Maria Bakardjieva, *Internet Society: The Internet in Everyday Life* (London: Sage, 2005).

15 Engin Isin, *Being Political: Genealogies of Citizenship* (Minneapolis: University of Minnesota Press, 2001).

16 Richa Nagar and Susan Geiger, "Reflexivity, Positionality and Identity in Feminist Fieldwork Revisited," in *Politics and Practice in Economic Geography*, ed. Adam Tickell, Eric Sheppard, Jamie Peck, and Trevor Barnes (London: Sage, 2007), 267–78.

17 Ananya Roy, "Why India Cannot Plan Its Cities: Informality, Insurgence and the Idiom of Urbanization," *Planning Theory* 8, 1 (2009): 76–87.

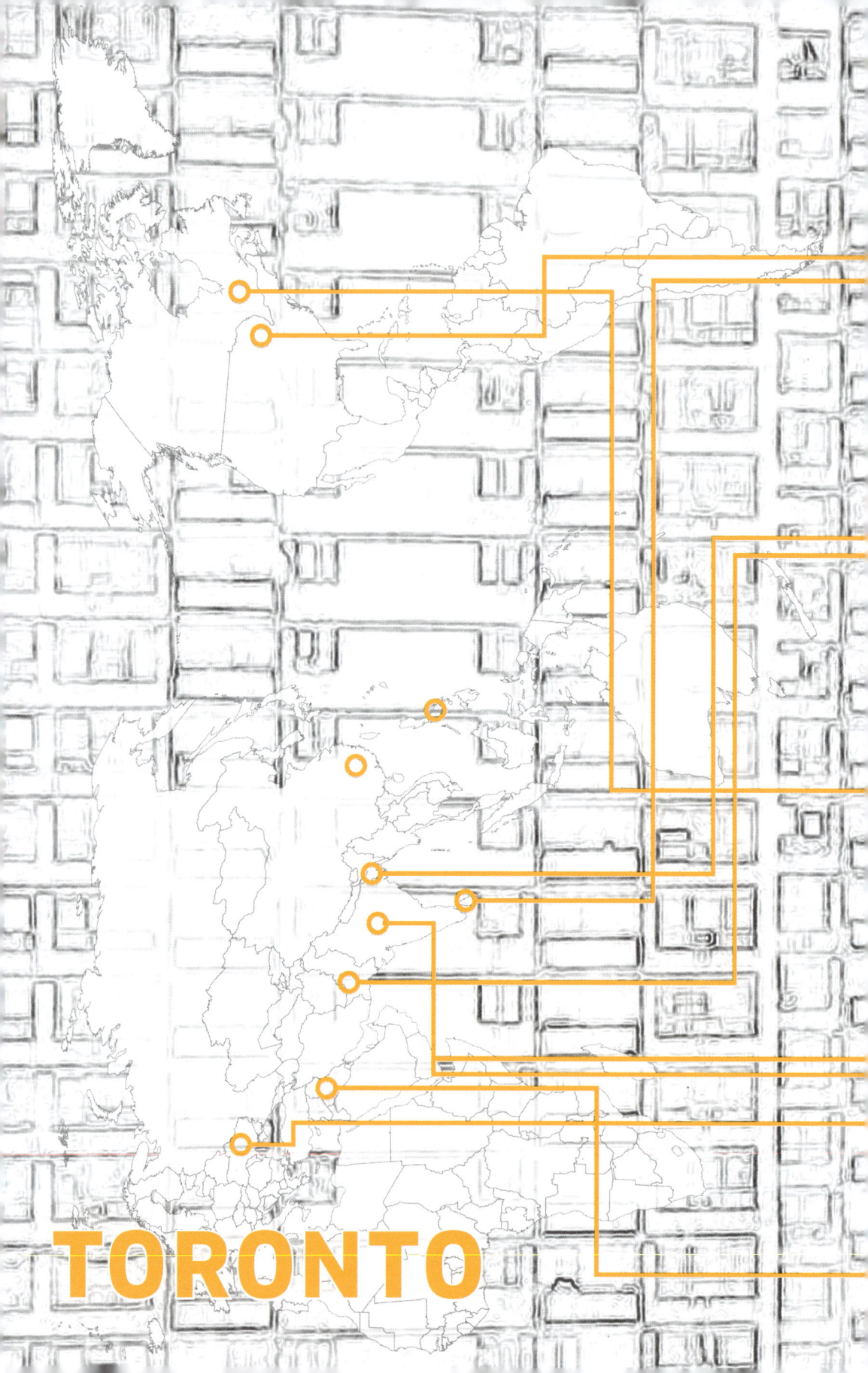

TORONTO

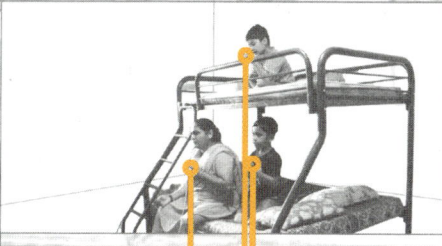

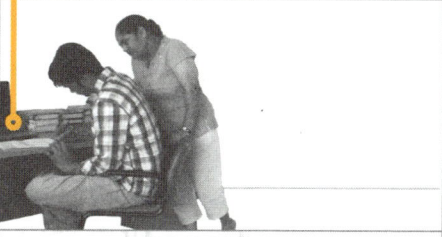

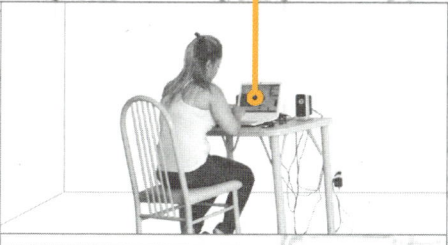

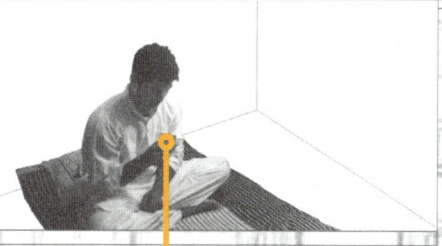

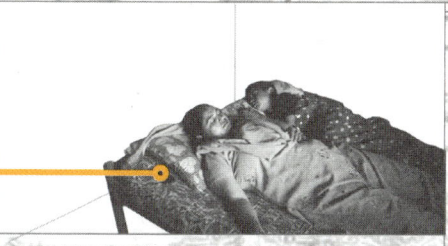

TORONTO

Digital Debt in a Precarious City

Emily Paradis, Heather Frise

IN THE LIVING ROOM of a ground-floor, mid-century apartment in Toronto, four women are talking about data.

> "The kids are addicted to it," one says.
> "True, but you need it to catch up with friends and family who live far away," says another.
> The third admits she feels empty and restless in bed without her phone – she needs music, Tumblr, and Facebook to soothe her to sleep.
> The fourth recounts that when she misplaced her phone in Grade 7, her parents refused to replace it. Now in Grade 12, she doesn't miss it – she connects with friends via Snapchat on her iPod.
> You need data, they all agree, and you want it. But it will cost you.
> "Pay as you go is terrible because you keep using it without knowing how much you're paying," one explains. "We had a 'complimentary' Rogers plan for one month, but the second month we got a bill for $125!"[1]

In a community where most households are living on about $2,500 a month, and with rent consuming almost half that amount, her companions understand the impact of an unexpected bill for $125.

The woman who said the kids are addicted commiserates. "When Rogers gave us the bundle, that's what messed me up. My daughter controlled her usage, but something made her gigabytes go up. They cut it off, and I am still paying for it for the life of the contract, even though she's not using it. They shouldn't be allowed to do that! I'm always balancing my money properly. I never thought I would be in this position."[2]

Stress is etched around her eyes. "I've been trying to get them to lower the bundle for a long time. I just want to cut off everything and just keep the land line. They're calling me twenty-four seven. I don't even want to look at the bill anymore. I just want to be at peace."

TOWERS IN A DIVIDED CITY

These women are talking about what we call "digital debt" – the financial and emotional cost of privatized digital technology for low-income consumers. Digital debt is emblematic of, and enables, a broader trend: the extraction of wealth from, and transfer of risk onto, low-income, racialized, and immigrant households and neighbourhoods in the inner suburbs of the neoliberal city.

Their conversation is taking place in a high-rise building on Kipling Avenue in Rexdale, a neighbourhood in the northwest corner of Toronto, Canada. The apartment is rented by a local agency. The kitchen cupboards are stocked with paper plates and plastic cutlery for community barbeques. The bedrooms hold donated office desks. Bulletin boards by the front door display leaflets for tenants' association meetings and neighbourhood services. Down the hall, off the building's main lobby, a large meeting room sits locked, the tape on the back of a fading community map slowly peeling off one of its tired beige walls. At one end of the room, computers donated by a local employment-skills organization stand unused in dusty carrels.

In the storeys above, and in the neighbouring building, 454 households share similar apartments: spacious but worn-out homes. Almost 1,200 such buildings can be found across Toronto, concentrated in the postwar inner suburbs. Built in the 1960s and '70s to house middle-income couples and young families working their way up to homeownership, these buildings were the product of federal programs that provided robust financing and incentives to private rental development and ample funding for construction of social housing.[3] The resulting "towers in the park" neighbourhoods are also products of provincial, regional, and municipal planning policies that shaped the development of the inner suburbs as mixed-income neighbourhoods with a range of housing forms and tenures, from rent-geared-to-income units in public-housing projects, to private rental apartments, to detached houses for rental or ownership.[4]

▲ **A typical mid-century apartment tower in Toronto.** Photograph by Jaime Hogge. Courtesy of the National Film Board of Canada.

▼ **The Rexdale building complex.** Photograph by Jaime Hogge. Courtesy of the National Film Board of Canada.

In the intervening decades, these neighbourhoods – and the city as a whole – underwent changes that mirrored neoliberal cities across North America and Europe.[5] Gentrification has brought upper-income professionals back into the detached old houses of the inner city, while new-build condominiums occupy the downtown's disused industrial sites and rail corridors, housing highly educated workers close to their jobs in the increasingly polarized knowledge economy. The aging rental towers of the inner suburbs, poorly maintained and no longer desirable, have filtered down.[6] They are now the landing place for households at the bottom of the labour and housing markets: racialized precarious workers, new immigrants and refugees, lone-parent families, low-income seniors, and people with disabilities. These deteriorating buildings and disinvested neighbourhoods are ill-equipped to meet the needs of their residents. Public transit is inadequate in these autodependent zones; services and amenities are scarce; the service-sector and manufacturing jobs on which most residents rely are located a great distance away; and the buildings' elevators, heating systems, common areas, waste-disposal infrastructures, and unit interiors are reaching the end of their lifespans.[7]

Elevators and other major systems in these towers are reaching the end of their lifespans.
Photograph by Jaime Hogge. Courtesy of the National Film Board of Canada.

Toronto's sociospatial transformations have produced a new urban landscape of deep inequality and segregation. While 66 percent of the city's neighbourhoods were middle income in 1970, only 29 percent were middle income in 2005.[8] Wealth and growth have consolidated in the city core and along the subway lines, where the average annual income is $88,400 and where 82 percent of residents are white (far above the city's average of 57 percent). Of the low-income neighbourhoods that make up 53 percent of Toronto's census tracts, most are in the inner suburbs, where incomes have steadily declined, averaging $26,900, and where 66 percent of residents identify as racialized.[9] When gender and other intersecting factors such as age are included in the picture, new sociospatial polarities come into view. There are forty-seven census tracts in which the average income for women aged twenty-five to sixty-four is less than half the city's average. In another census tract, the income of working-age men exceeds Toronto's average by a factor of thirteen.[10]

LANDSCAPES OF DEBT, LANDSCAPES OF RISK

As urban geographer Alan Walks shows in his research and contribution to this book, this divided urban landscape is also a debtscape. Debt, Walks notes, is a key feature of the neoliberal city: it is both an enabler of the financialization driving urban economic growth and development and a fact of everyday life for most city dwellers.[11] Between 1984 and 2009, during the same period in which economic inequality widened dramatically in Canada's cities, levels of household debt doubled.[12] Even the global financial crisis of 2008 and its aftermath failed to slow the growth in debt as a proportion of disposable income in Canada. The bulk of household debt in Toronto is mortgage debt, which has been driven by feverish speculation in Toronto's urban land market. But unsecured consumer debt – from credit cards, student loans, car purchases, and other kinds of registered loans – has also increased to what many commentators consider dangerous proportions: in 2009, the average Toronto household owed forty-seven cents in nonmortgage, non-credit card debt for every dollar in after-tax income.[13]

Like income and wealth, debt is unevenly distributed across the population. The majority of indebted households have annual incomes below $50,000. Younger adults and lone-parent households have the highest debt-to-income ratios. Low-income households, women, and racialized

Condominiums, as seen from an apartment window. Photograph by Jaime Hogge. Courtesy of the National Film Board of Canada.

groups carry the highest interest rates and the greatest revolving monthly debt.[14] Warnings about Canadians' debt-fuelled lifestyles can often contain moralistic overtones. Research offers a more nuanced understanding of the role of credit in the neoliberal economy, particularly for low-income groups. A 2011 survey found that 57 percent of indebted Canadians cited daily expenses as the primary reason for their use of credit.[15] In their research on indebted households in Vancouver, Scott Graham, Emily Hawes, and David Ley found that "lower-income borrowers, single-parent families, young borrowers, and recent immigrants tend to accumulate debt in relation to structural disadvantages (e.g., underemployment, low wages and high cost of living), strong marketing of credit products, and interruptions to income."[16]

Debt is also unevenly distributed across space. Mortgage debt and overall rates of debt are highest in the gentrified city-core neighbourhoods, where prices for detached houses and new condominiums continue to spiral upward. But debt as a proportion of income is highest in the suburbs and inner suburbs, and unsecured consumer debt is highest in low-income neighbourhoods.[17]

Walks suggests that debt, with its regressive social and spatial distribution, is an important mechanism for the transfer of wealth between populations and neighbourhoods: from young people to older adults and from impoverished inner suburbs to the wealthy enclaves of the city core. It is also, he underlines, a mechanism for the transfer of risk: "It is those immigrant-reception neighbourhoods concentrating multi-family households and visible minorities that have higher levels of indebtedness, suggesting that racialized immigrants are disproportionately bearing the risks of global city evolution under financialization."[18]

Indeed, risk itself is a feature of the neoliberal city: urban economic growth relies on speculation and leverage. This risk is displaced onto groups and places at the margins of the city through policy and market mechanisms that interlock to produce precarity for individuals, households, and neighbourhoods. Federal immigration policy, for example, has changed course from fostering population growth through the permanent settlement of families to providing temporary migrant workers – mainly from the global south – to labour-market sectors, including agriculture, construction, and services.[19]

Meanwhile, those settling permanently in Canada face lengthening periods of precarious status during which they lack the full social entitlements of permanent residency, leading to deep and long-lasting income disparities between Canadian-born workers and those born elsewhere.[20] In the polarized labour market of the knowledge and service economies, an increasing proportion of workers have precarious jobs, and women and racialized groups are concentrated in the lowest-paying sectors.[21] Elimination of rent control, termination of state social-housing programs, and speculation-driven housing development have produced a polarized housing system in which an increasing proportion of renters belong to the lowest income segments of the population.[22] Gentrification, displacement, and discrimination force low-income immigrant and racialized families into the deteriorating rental stock of the inner suburbs, where units are unaffordable, overcrowded, and in disrepair; tenancies are precarious; and rates of hidden homelessness are high.[23]

Rather than providing a measure of financial security in tough times, social programs such as welfare and disability benefits instead maintain recipients in a state of perpetual insecurity, channelling them into the most marginal sectors of the labour market.[24] As Krystle Maki's research and contribution to this book show, technological changes in the delivery of these programs have facilitated the disentitlement of recipients through surveillance protocols that quantify their risk of contravening eligibility criteria.[25] Social infrastructures, too, are increasingly unreliable: funding for social programs comes and goes; amenities and services are hard to reach; the bus is never there when you need it. Risk and precarity at every level, then, characterize daily life at the social and spatial margins of the city.

It is well-established that the digital is central to these social and spatial changes. Digital technologies are not only key commodities of urban consumer culture and the driving technologies of the knowledge and service industries – they also directly, materially enable the financialization and speculation fuelling the neoliberal city's economic growth.[26] Digital technologies are also increasingly critical to the urban social movements that are emerging to contest the global ascendancy of capital and its racialized and gendered mechanisms of coercion and surveillance – even as they extend the reach of that surveillance and facilitate the cultivation of neoliberal subjectivities.[27] What has received less attention, though, is the direct role of digital technologies in the extraction of wealth from, and the transfer of risk onto, individuals, households, and neighbourhoods on the social and spatial margins of cities like Toronto.

AS ESSENTIAL AS WATER

Our investigation began in 2011, when our team of scholars and documentary filmmakers worked with residents of the Rexdale buildings to conduct a survey on access to and use of information and communication technologies (ICTs) in their apartment complex. The survey was one component of a years-long engagement between the National Film Board's *Highrise* team, the building's tenants' association, and a local community-development organization. This evolving collaboration also included a Photovoice project, in which residents documented their homes and daily lives, and a charrette, which brought together residents, architects, designers, and animators to reimagine the open space around the buildings. This community-based, participatory approach aimed to mobilize research and documentary in support of community development and resident self-advocacy. The process yielded two award-winning films and catalyzed action and relationships within and outside the building.[28]

For the survey, tenants who were trained in research methods went door to door interviewing their neighbours about their digital lives. The community-based researchers commented that they still didn't know their neighbours, in spite of two years of organizing by the Action for Neighbourhood Change program and a newly formed tenants' association. Crossing the boundary into one another's homes introduced a new depth to residents' connections.

The survey also generated evidence to support the tenants' association's neighbourhood improvement projects: for example, survey data on the number of children in the building were the basis for a successful funding application to build a new playground on the site. The documentaries, meanwhile, provided a platform for the residents' advocacy: artwork from the project was featured in a poster campaign on the city's transit system; an influential morning radio show held a live broadcast from the building's lobby; one of the films was launched at a public screening in the rotunda of Toronto's City Hall; and leaders from the tenants' association met with the mayor.

Although the literature on the digital divide in low-income, racialized, immigrant, and suburban communities suggests that these groups have low rates of access to the internet and digital technologies, our team's prior experience in this neighbourhood and others like it had alerted us to the paradoxical ubiquity of technology in communities that seem least able to afford it. The survey corroborated these observations: in 2011, 80 percent of respondents had internet access at home; when the survey was repeated in 2014, the rate had risen above 90 percent. At both points in time, rates of access in this low-income community were equal to or greater than the Canadian average. Beneath this simple statistic about home-based access, though, lay different patterns of use, ranging from residents who limit time online to a few minutes a day checking email and news to super-users who spend whole days on Skype with family members around the globe.

We also learned that residents were using a broad range of technologies: tablets, hand-held devices, gaming systems, desktop computers, laptops, land lines, and mobile phones. With limited access to costly hardware, many were making creative use of whatever technology they had available. One young member of our research team, for example, was accessing email and internet through his PlayStation – the only internet-enabled device in his home.

In surveys and focus groups, residents cited a wide range of uses for ICTs: maintaining connections with friends and family members in the

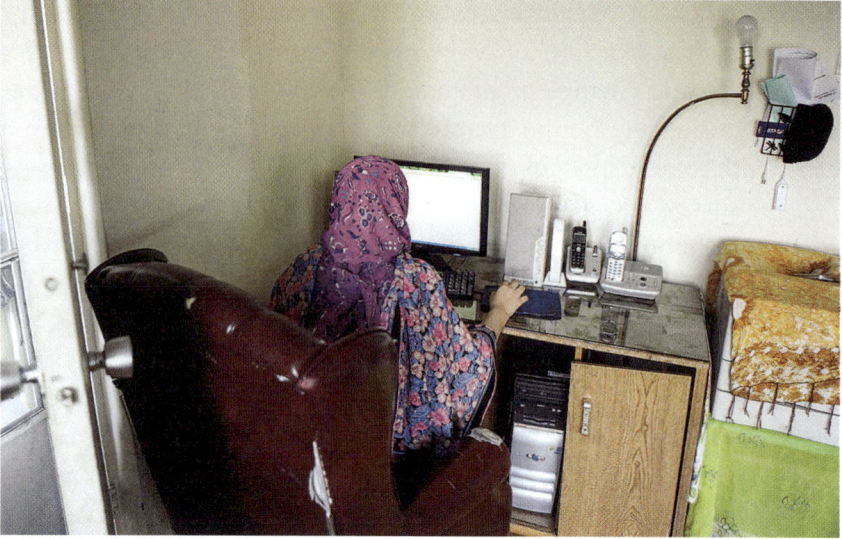

▲ **ICTs, a critical infrastructure for residents.**
Photograph by Jaime Hogge. Courtesy of the
National Film Board of Canada.

▼ **The importance of ICTs to children's education.**
Photograph by Jaime Hogge. Courtesy of the
National Film Board of Canada.

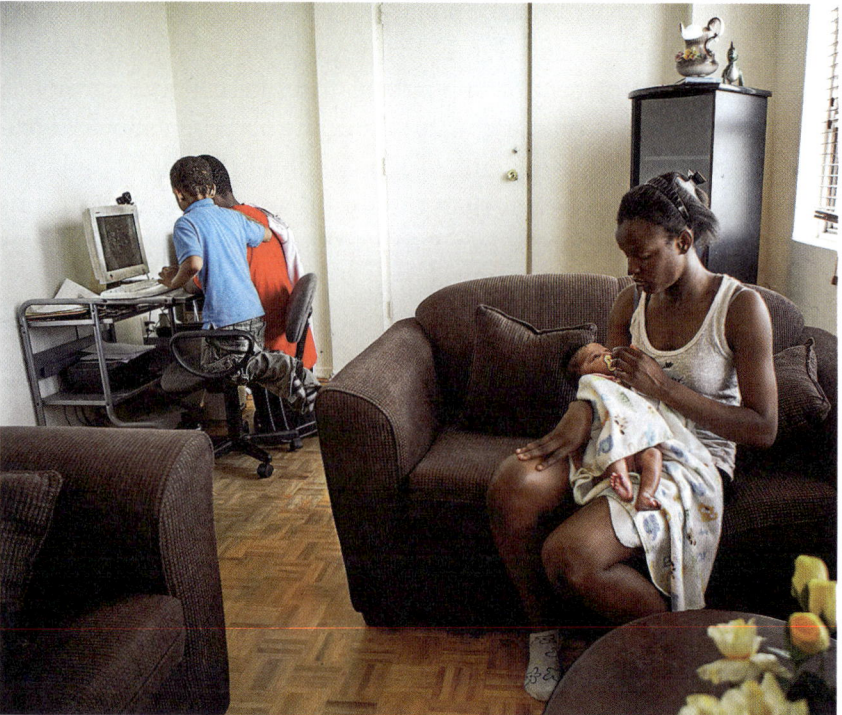

building and across the globe, completing children's school work, obtaining medical and health information, connecting to faith communities, accessing government information, researching employment opportunities, pursuing online courses, shopping, and enjoying culture and entertainment. The use of social-media platforms such as Twitter, WhatsApp, Facebook, Tumblr, and Snapchat was high, and it increased from 2011 to 2014. Many respondents reported that their most important reason for having digital services was to provide their children with the tools they required for school success. Within the household, the economies of ICT use were sometimes stratified by age and gender – for instance, male breadwinners and students would be accorded priority access over female household members – but this was less common than anticipated. Children tended to be the most frequent and most skilled users of ICTs, acting as translators of the online world for their parents.

Whereas conventional ideas of a digital divide tend to imply that low-income and newcomer communities lack technological literacy, we found the opposite. The residents' global connections alerted them to platforms and trends that had yet to hit the North American mainstream. For example, many were using WhatsApp in 2011, long before it was bought by Facebook and popularized in North America. In focus groups and community meetings, residents schooled the research team on new apps and devices, such as Gogo and magicJack, that provided improved access and reduced cost. At the same time, some groups – notably older women – had little facility with the devices in their homes.[29]

Residents repeatedly underlined the absolute necessity of ICTs in their daily lives. Without these technologies, they would lose access to vital connections and indispensable information. As one resident put it, the internet was "like water."

While these findings belie the myth of a digital divide in home-based access, they underscore an emergent divide in the workplace. In contrast with the ubiquity of connection in the jobs of the "creative economy," very few residents had access to the internet at work – but this is not to say that ICTs were absent from their waged labour. They were, in fact, omnipresent in structuring and regulating residents' workplaces and workdays. For example, many residents held jobs in the warehouses and distribution centres surrounding the airport in an adjacent suburb. These workplaces are highly digitized and securitized spaces where workers surrender their devices on entry and submit to continuous monitoring by closed-circuit cameras feeding real-time data to a security hub. Meanwhile, service-sector work in franchise restaurants – also common among residents – is increasingly scheduled by centralized systems that use data on patterns of customer visits to allocate employees to on-call shifts, sometimes across multiple locations. These just-in-time arrangements carve workdays into low-paying segments separated by long commutes. They give workers and site managers little discretion to adapt schedules to meet child-care needs and other obligations.[30]

INFRASTRUCTURES OF CITIZENSHIP

The Rexdale rental complex, like many others across Toronto, is what Doug Saunders calls an arrival city.[31] It is home to economic immigrants from the Caribbean, Africa, South Asia, and West Asia and to successive clusters of refugees displaced from sites of violence, crisis, and disaster, including Syria, Iraq, Haiti, and Afghanistan. Turnover is high, as many residents move on to other homes and neighbourhoods once they become established. More than 90 percent of survey respondents were born outside Canada, and almost half had been in the country for less than five years. For them, digital technologies were necessary tools in their navigation of transnational lives and in their efforts to hold together intimate ties stretched around the globe.

For these reasons, many explained, obtaining digital services was one of their top priorities on arrival in Canada. Several noted that after finding a home, their next step was to find a phone and internet plan. Some, in fact, obtained their first Canadian phone at the airport. One young woman our team met in 2011 was living in an apartment that was completely empty except for a bed, a table, a chair, and a laptop. A refugee from Iraq, "Angel," who had arrived in Canada only a short time before, had obtained an internet connection immediately on finding an apartment. Completely alone in Canada, she spent hours each day on Skype, on email with her fiancé and family, and on news sites, where she witnessed catastrophe unfold in her home region. Though she physically resided in a Rexdale apartment, she was living in an in-between virtual space in connection with loved ones still in Syria and Iraq. The internet was her lifeline: she commented that without it, she would die.[32]

Like water, ICTs were not only necessary to residents' daily lives – they also enabled ongoing growth and transformation. In this diasporic

community, we learned, the internet is a critical means of cultural, religious, and political participation, forging new connections within and across national borders and new meanings of citizenship. For example, the Patels, another family we met in 2011, meditated daily on sacred images of Hindu deities posted online from cities around the world, and they proudly shared online photos of their home temple in a nearby suburb, which Ms. Patel had helped to build.[33] Another family, a couple in their eighties who had been in Canada since 1965, used online news and community websites to monitor the political intrigue surrounding the completion of a new bridge in their rural Jamaican hometown. These and other developments in their country of origin sparked conversation during weekly visits to their neighbourhood seniors' club. Another resident, a refugee claimant from Nigeria still awaiting permanent status in Canada, introduced her children to Nollywood movies on Netflix. After a hard week of work and school, this Sunday after-church ritual was, for her, a time to relax as a family and foster the kids' connection to Nigerian culture and identity. Finally, many residents, when asked about their principal news source, cited websites from their places of origin. Through GhanaWeb, Television Jamaica, Nigeriaworld, and Al Arabiya, residents engaged with events and perspectives that received little notice in mainstream Canadian media.

ICTs thus facilitate and make visible residents' global webs of affiliation, unsettling simplistic narratives of immigrants' aspiration to integrate into a monolithic Canadian identity. But as is illustrated by Angel's story, in a world where migration is so often a product of economic dispossession and violent displacement, the digital is also a realm of yearning, nostalgia, and grief through which residents struggle to hold fast to distant loved ones or watch helplessly as disaster consumes their homelands. One respondent, for example, worried that her husband was not settling in Canada, as he spent all his time on news sites from back home.

The geography of the residents' connections also complicate understandings of urban citizenship. When asked to locate their five most frequent online contacts, residents overwhelmingly named contacts half a world away and contacts very close by – in the building or the immediate neighbourhood. Their online connections, and their daily travels, also extended outside of or across Toronto, from suburb to suburb. But not one resident named a main contact in downtown Toronto, and few residents ventured downtown on a regular basis. Some had never, in their time in Canada, made the twenty-four-kilometre trip downtown (an hour by car or two or more hours by transit). Though Toronto prides itself on being a city of neighbourhoods, the residents' physical and virtual connections traced a more fragmented map of the city, one traversed by thick intra-suburban networks and only faint links to the city's self-declared geographic, economic, and social centre.

Accounts of ICT use and access in this inner-suburban neighbourhood, then, call into question and complicate simplistic understandings of the digital divide as an exclusively income-based discrepancy in material access to technology. Instead, what their experiences revealed was complex interplay between access and exclusion at multiple scales. At home, rates of access and use signalled the residents' vigorous and creative uptake of digital technologies by any means necessary. At work, ICTs both signified and reproduced a digital (class) divide: on the one side, those who can use ICTs to determine the time and place of their labour; on the other, those whose employers use ICTs to constrain their choices and surveil them. In this diasporic community, ICTs work through and extend spatial and emotional processes of settlement, enabling the development of hybrid identities and transnational citizenships, even as they stretch spirits to the breaking point. The residents' everyday geographies suggest an entanglement of virtual and physical mobilities, in which ICTs afford access to people and places near and far while reinscribing the city's deep social and spatial divides.

THE HIGH PRICE OF PRIVATE INFRASTRUCTURE

Though it is, without doubt, a critical infrastructure of citizenship, internet provision has little state involvement in Canada. Digital services often rely on publicly funded infrastructures for their transmission but are delivered exclusively through the private market. Although traditional telecommunications, including broadcast television, radio, and telephone, have long been regulated in Canada, private companies that provide digital services have enjoyed a long period during which they faced little state regulation. As a result, costs for digital services in Canada are among the highest in the world.[34]

As the women whose conversation opens this chapter would attest, in this community of low- and moderate-income households, the high cost of digital services is a significant concern. In our 2014 survey, for example, most respondents reported monthly incomes in the range of $2,000 to $3,000. The costs of internet and mobile phone access were more than $100 a month for most and more than $200 for many.[35] Some households were spending 10 percent of their already low monthly income on digital services. At the same time, many were spending 50 percent or more of their income on rent. The aging infrastructure of their buildings, though, meant that they were paying high prices for service of lower quality. High-speed data transmission was compromised by outdated wiring, and concrete-block walls made wireless access spotty and erratic.

Tenants of Toronto's major social-housing provider are eligible for reduced internet service rates and subsidized computers.[36] But measures have not been taken to make ICT access affordable for low-income households in privately owned rental buildings, where the majority of low-income people in Toronto live. Their high rent expenses make it even more difficult for them to cover digital-services costs.

Like rent and other utilities, digital access was considered a fixed cost by most of the households we interviewed – something they couldn't do without. In the context of low wages, precarious employment, and high rent, most residents were sacrificing other items they considered discretionary to afford the high price of digital services. They reported that they must sometimes sacrifice personal items (75 percent), recreational activities (53 percent), and even basic needs such as groceries (41 percent) to afford these services. One single mother, for example, explained that she managed to squeeze internet into her tight monthly budget by discouraging what she referred to as unnecessary eating, such as after-school snacks. She reasoned that her children's academic success was worth afternoon hunger pangs.

MECHANISMS OF EXTRACTION AND RISK

This account is a visceral example of how privatized digital infrastructure enables extraction of resources from low-income bodies, households, and neighbourhoods. But our interviews and focus groups revealed more

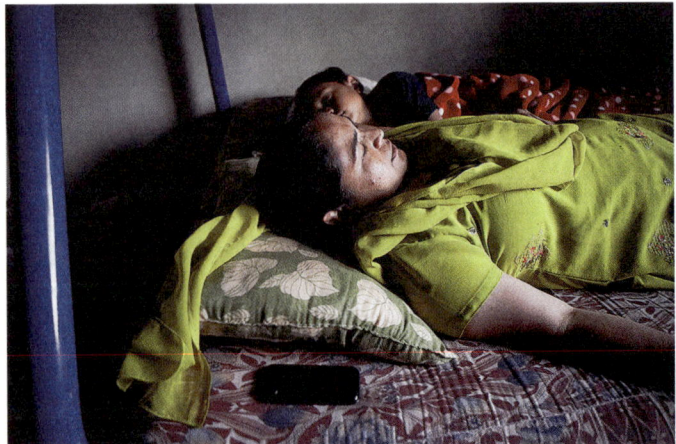

The Patel family fall asleep to chanting. Photograph by Jaime Hogge. Courtesy of the National Film Board of Canada.

complex dimensions of this dynamic at work, as opaque delivery mechanisms, inscrutable billing structures, regulatory gaps, and high-pressure marketing campaigns converge to draw low-income consumers into digital debt. As the conversation among the four women illustrates, lack of control and knowledge are key themes in the residents' accounts of digital debt. For households that carefully monitor every expense, purchasing digital services is an unpredictable gamble. As one resident explained, "Before I had unlimited data, I had just internet from [provider]. Sometimes, I would ask them, 'What do you think I will pay by the end of the month?' They would say, 'Fifty dollars. With tax, maybe $60.' But eventually, when I see the bill, it's $110! ... They have a way of increasing the bill."

One way to increase the bill is overage – additional charges levied when users exceed their allotted minutes or data. Phone minutes are difficult for consumers to calculate, and until recently data usage has been impossible to monitor.[37] Some measures have recently been taken by state regulators to limit the overage that can be charged and to require providers to alert customers once they have gone over their limit.[38] But even once alerts are sent, minutes and data remain available to users – at a premium price – and the onus is on them to shut down their devices or refrain from accessing the internet.

Contracts, too, pose particular difficulties for households with precarious incomes. Because they are locked in for a fixed period, consumers are prevented from taking advantage of better deals. Cancellation comes at a high price, and unpaid cancellation fees affect consumer credit ratings. One respondent who was in her early twenties talked about renting her first apartment with her boyfriend. The cost of digital services plunged them into debt. She reported, "My bill was going higher and higher, up to $300, so I cut it off, but then I had to pay an additional fee for early cancellation. I wanted to pay it, but at the time, financially, we were going through a lot. I didn't get a stable job until recently. [The provider] was chasing me. The interest was going up. They had a collection agency after me. It was affecting my credit rating, so when I got my tax money, I used it to pay off my debt."

Residents also reported being taken advantage of by offers of low introductory rates that then expired after a brief time, leaving them tied to a contract they couldn't afford. A senior living on a fixed income reported: "They tell you about a new package deal. For the first three months it's free or at a discount, but then it goes right back up to top price. When you tell them you want to cancel, they say you have a contract for two years and if you want to get out of it you will have to pay."

Several respondents shared stories of calling their provider in a panic on receiving an unexpectedly high bill and pleading for amnesty. In these instances, the providers rarely forgave an amount owing, even if the overage was unintentional. Instead, residents explained, "help" with these situations usually came with new costs, such as a payment plan or an obligation to renew or extend their contract.

The residents' stories echo accounts of other predatory schemes to siphon capital out of low-income, racialized households and neighbourhoods. The most thoroughly documented is subprime mortgage lending in the United States, which precipitated the global financial crisis and caused hundreds of thousands of households – disproportionately low-income and African American – to lose their homes. The payday lending industry is also known to concentrate its storefronts in proximity to low- and moderate-income neighbourhoods.[39] These schemes not only target the same groups of marginalized consumers, they employ similar tools: introductory teaser rates that increase after a brief period; hidden fees and byzantine formulas that inflate amounts owing; contracts that prevent consumers from extricating themselves; and compounding penalties that induce inescapable cycles of mounting debt.[40]

Excluded from formal markets, consumers at the margins of the divided city must seek work, housing, and services in unregulated, informal, and substandard markets in which they pay more for products of lower quality and face greater risks in doing so. For example, discrimination and other barriers force lone mothers into sectors of the housing market where they pay higher-than-average rent for the most precarious housing in the worst condition.[41] Likewise, digital services that appear to be the least expensive and the most accessible to those with poor credit histories and insecure incomes carry the greatest risk of increased costs. By contrast, more expensive plans guarantee predictability. The gap is not only economic and social, it is also spatial: while workers in Toronto's downtown knowledge economy enjoy ubiquitous digital access at work and in public spaces, low-income, diasporic suburban residents must pay for virtual mobility to bridge global distances, remedy geographic isolation, and reclaim time spent in long commutes.

If credit itself is a necessary infrastructure of the new economy, then communities at the social and spatial edges of the city encounter unevenness, insecurity, and risk when they seek to access it. In fact, rather than facilitating the workings of daily life, as infrastructures usually do, credit has become a mechanism of predation, extracting resources from and diminishing the quality of life of low-income families. Digital technologies are a critical enabler of this extraction because, as Safiya Umoja Noble and Brendesha M. Tynes point out in *The Intersectional Internet*, "Information, records, and evidence can have greater consequences for those who are marginalized."[42]

Consider the processes by which borrowers with the lowest income come to pay the highest interest rates. Banks have retreated from low-income neighbourhoods, and their requirements for loans often exclude marginalized consumers. The new digital mechanisms of surveillance now make bank accounts visible to creditors, which means a deposit could trigger calls from debt collectors.[43] Consumers at the margins – excluded from banks by geography, social status, or the risk of garnishment – are forced to obtain financial services from payday loan centres, where their debt loads continue to multiply unseen. The benefits of this arrangement flow back to banks at the city's centre because many payday loan operations are owned by banks.[44]

"THEY SHOULDN'T BE ALLOWED TO DO THAT"

As wages have fallen or stagnated and work has become increasingly precarious, credit has become a way for low- and moderate-income households to bridge the gap and meet daily needs.[45] In the context of fluctuating incomes, households use debt to manage risk and cover fixed costs such as rent and child care. In spite of these causes, high levels of indebtedness in Canada continue to be framed – whether by the Bank of Canada, media commentators, or credit counselling agencies – as a moral issue. Debt, they argue, is an individual responsibility, and high debts indicate a lack of discipline.[46]

All of this risk and responsibility is a lot to bear, and it exacts a toll that is not only financial but also emotional. Debt stress includes feelings of fear, shame, isolation, desperation, and loss of hope, regardless of whether the debt was accrued to pay for necessities or nonessentials.[47]

Our interviews corroborated this. Residents blamed themselves for accumulating digital debt. In addition, the privatization of digital technologies and their representation as commodities whose purpose is entertainment obscured the digital's status as an essential infrastructure.

In *Disorderly People,* Joe Hermer and Janet E. Mosher suggest that neoconservative policy reforms that mobilize moral discourses are "intentionally designed to dismantle both the material and emotional infrastructure of the welfare state."[48] Indeed, subjectivities of responsibilization are a crucial component of neoliberal hegemony. In other words, in an economic and social system in which everyone is considered a radically free agent, if we can't secure the necessities of life, we have only ourselves to blame.

Yet the residents we worked with resisted these attempts to render them responsible. In meetings and focus groups, they shared information about devices and hacks that could be used to obtain services at reduced costs, techniques to avoid penalties, and effective arguments for wringing concessions from digital service providers. They countered representations of the digital as frivolous entertainment, insisting upon its status as a necessary infrastructure, as necessary as water, one that the state should make available to citizens as a social right.

The residents' responses to digital debt opened a path to broader critiques of prevailing economic, social, and spatial arrangements, and these claims extend beyond the digital realm. Like protesters who wave banners that read "We won't pay for your crisis," the residents refused, in both material and discursive ways, to carry the burden of risk associated with the global city's evolution under financialization.

As Margit Mayer has suggested, the discourses, tactics, and social movements that are challenging the deep inequities of the current system are emerging from the margins of the neoliberal city.[49] To ascribe systems-toppling power to these nascent instances of agency would be to romanticize them.[50] But in them new spaces and forms of resistance are coming into view. They are anchored in hyperlocal places and globally networked. They are led by racialized migrants, many of them women, and concerned with the stuff of the everyday – rent, roaches, bills, the bus. They are converging not in the workplace but in the neighbourhood. And they are skilfully deploying homemade hacks to circumvent predatory traps while amassing the cache of tools required to navigate the neoliberal global city.

A picnic sponsored by the residents' association. Photograph by Jaime Hogge. Courtesy of the National Film Board of Canada.

● ● ● ● ●

Back on the ground floor of the building on Kipling Avenue, the tenants' association is meeting. Members enumerate problems with the building's infrastructure, recount repeated episodes of disrespectful treatment from the building's management, and take stock of the small but tangible improvements they have won over the course of more than five years' hard work: functioning elevators, a community garden, and a playground for the kids. Still, rents continue to rise while conditions deteriorate. There is so much more to be done, and the City program that funded the establishment of the tenants' association has just been terminated. The frustration is palpable. Some say they are ready to give up.

The woman whose daughter's phone was cut off leans forward. She is a leader here, and people listen when she speaks. She looks around the table and gives voice to the question in everyone's mind. "Why should we have to do the dirty work," she asks, "when they get to have their cake and eat it too?"

NOTES

1 Canada's telecommunications sector is heavily concentrated. Five huge companies – Rogers, Bell, Telus, Shaw, and MTS/Allstream – reap 84 percent of revenues from telecommunications services. Canadian Radio-television and Telecommunications Commission, "Communications Monitoring Report 2015: Telecommunications Sector Overview," http://www.crtc.gc.ca/eng/publications/reports/policymonitoring/2015/cmr5.htm.

2 Digital services providers aggressively market bundles through which customers receive a discounted monthly rate in exchange for purchasing data, cable television, telephone, and mobile phone services from the same company for a contracted period of time.

3 Greg Suttor, "Rental Market and Policy Comparison: Toronto, Montréal and Vancouver" (presentation of work in progress, Neighbourhood Change Research Partnership Research Day, Toronto, May 7, 2015), on file with author.

4 Glen Searle and Pierre Filion, "Planning Context and Urban Intensification Outcomes: Sydney versus Toronto," *Urban Studies* 48, 7 (2011): 1426.

5 Margit Mayer, "First World Urban Activism," *City: Analysis of Urban Trends, Culture, Theory, Policy, Action* 17, 1 (2013): 5–19.

6 Greg Suttor, "Rental Housing Dynamics and Lower-Income Neighbourhoods in Canada," Research Paper 235, Neighbourhood Change Research Partnership, Cities Centre Research Paper Series, University of Toronto, 2015.

7 United Way Toronto, *Poverty by Postal Code 2: Vertical Poverty, Declining Income, Housing Quality and Community Life in Toronto's Suburbs* (Toronto: United Way Toronto, 2011).

8 J. David Hulchanski, *The Three Cities within Toronto: Income Polarization among Toronto's Neighbourhoods, 1970–2015* (Toronto: Cities Centre, University of Toronto, 2010).

9 Ibid. The data here, drawn from Hulchanski's 2010 study, *The Three Cities within Toronto*, propelled the issue of sociospatial inequality in Toronto into public and political debate. While reliable updates are available for the income statistics via tax-filer data, Canada's national five-year census was replaced in 2011 by a voluntary survey. As a result, reliable data on race, immigrant status, and other variables are not available for 2011. The mandatory long-form census was reinstated in 2016.

10 Custom analysis of 2012 Canada Revenue Agency data by Richard Maaranen of the Neighbourhood Change Research Partnership, Factor-Inwentash Faculty of Social Work, University of Toronto (principal investigator: J. David Hulchanski).

11 Alan Walks, "Mapping the Urban Debtscape: The Geography of Household Debt in Canadian Cities," *Urban Geography* 34, 2 (2013): 153–87.

12 Scott Graham, Emily Hawes, and David Ley, *Metro Vancouver's Debtscape* (Vancouver: Social Planning and Research Council of British Columbia, 2016), 18.

13 Walks, "Mapping," 167.

14 Graham, Hawes, and Ley, *Metro Vancouver's Debtscape*, 18.

15 Certified General Accountants Association of Canada, cited in Walks, "Mapping," 160.

16 Graham, Hawes, and Ley, *Metro Vancouver's Debtscape*, 41.

17 Walks, "Mapping," 180.

18 Ibid., 180.

19 Nandita Sharma, *Home Economics: Nationalism and the Making of "Migrant Workers" in Canada* (Toronto: University of Toronto Press, 2006).

20 Luin Goldring and Patricia Landolt, *The Impact of Precarious Legal Status on Immigrants' Economic Outcomes*, IRPP Study No. 35 (Montreal: Institute for Research on Public Policy, 2012).

21 Poverty and Employment Precarity in Southern Ontario (PEPSO) Research Alliance, *It's More Than Poverty: Employment Precarity and Household Well-Being* (Hamilton, ON: McMaster University, 2012); and Sheila Block and Grace-Edward Galabuzi, *Canada's Colour-Coded Labour Market: The Gap for Racialized Workers* (Toronto: Wellesley Institute and Canadian Centre for Policy Alternatives, 2011).

22 Suttor, "Rental Housing Dynamics."

23 Emily Paradis, Ruth Wilson, and Jennifer Logan, "Nowhere Else to Go: Inadequate Housing and Risk of Homelessness among Families in Toronto's Aging Rental Buildings," Cities Centre Research Paper 231, Cities Centre, University of Toronto, 2014.

24 Jamie Peck, *Workfare States* (New York: Guildford Press, 2001).

25 Krystle Maki, "Neoliberal Deviants and Surveillance: Welfare Recipients under the Watchful Eye of Ontario Works," *Surveillance and Society* 9, 1–2 (2011): 47–63.

26 Saskia Sassen, "Mortgage Capital and Its Particularities: A New Frontier for Global Finance," *Journal of International Affairs* 62, 1 (2008): 187–212.

27 Margit Mayer and Jenny Kunkel, "Introduction," in *Neoliberal Urbanism and Its Contestations: Crossing Theoretical Boundaries*, ed. Jenny Kunkel and Margit Mayer (London: Palgrave Macmillan, 2012), 3–26; Brendesha M. Tynes, Joshua Schuschke, and Safiya Umoja Noble, "Digital Intersectionality Theory and the #BlackLivesMatter Movement," in *The Intersectional Internet: Race, Sex, Class, and Culture Online*, ed. Safiya Umoja Noble and Brendesha M. Tynes (New York: Peter Lang, 2016), 21–40; and Engin Isin and Evelyn Ruppert, *Being Digital Citizens* (London: Rowman and Littlefield, 2015).

28 See *The Thousandth Tower: Stories from Inside a Toronto Suburban Highrise*, http://highrise.nfb.ca/thousandthtower/, and *One Millionth Tower*, http://

highrise.nfb.ca/onemillionthtower/1mt_webgl.php, directed by Katarina Cizek.

29 In response to this identified need, the research team worked with the tenants' association to offer multilingual workshops in the buildings' shared meeting room, in which residents set up Gmail accounts and learned to use Skype.

30 David Friend, "Precarious Employment: Why 'On-Call' Shifts Have Workers Stressed, Activists Fuming," *CTV News*, September 3, 2015; and Seres Lu, "On-Call Scheduling under Increasing Scrutiny in Canada," *Globe and Mail*, September 6, 2015.

31 Doug Saunders, *Arrival City: The Final Migration and Our Next World* (Toronto: Vintage, 2010).

32 Angel's story is documented in the National Film Board documentary *Highrise*.

33 The Patels are also featured in *Highrise*.

34 Christine Dobby, "How Canada's Internet, Wireless Rates Compare with International Prices," *Globe and Mail*, August 11, 2016.

35 This cost range is similar to that being paid by Canadian households in general, according to a recent report. See John Lawford and Alysia Lau, *No Consumer Left Behind: A Canadian Affordability Framework for Communications Services in a Digital Age* (Vancouver: Public Interest Advocacy Centre, 2016).

36 Laurie Monsebraaten, "Anti-poverty Advocates Call for Affordable Internet," *Toronto Star*, February 6, 2016.

37 Major providers now enable customers to access information about their data usage online.

38 It is worth noting that regulatory changes are typically in response to issues affecting mainly middle- and upper-income consumers. An example is the curtailment of excessive charges and the imposition of alert systems for data roaming when service users leave the geographic area covered in their plan. These changes were made in response to public outrage spurred by media accounts – and first-hand experiences – of travellers returning from international vacations or business trips to bills for thousands of dollars.

39 Graham, Hawyes, and Ley, *Metro Vancouver's Debtscape*, and Walks, "Mapping."

40 After the research was completed, a public inquiry by Canada's telecom regulator vindicated these concerns. ACORN Canada (see Judy Duncan's essay in this volume) was a leading participant. See Public Interest Advocacy Centre, "CRTC Report on Telecom Sales and Practices Vindicates Consumer Concerns," https://www.piac.ca/our-specialities/crtc-report-on-telecom-sales-practices-vindicates-consumer-concerns/.

41 Maureen Callaghan, Leilani Farha, and Bruce Porter, *Women and Housing in Canada: Barriers to Equality* (Toronto: Centre for Equality Rights in Accommodation, Women's Housing Program, 2002).

42 Safiya Umoja Noble and Brendesha M. Tynes, "Introduction," in *The Intersectional Internet: Race, Sex, Class, and Culture Online,* ed. Safiya Umoja Noble and Brendesha M. Tynes (New York: Peter Lang, 2016), 3.

43 John Stapleton, *Welcome to the Financial Mainstream? The Hazards Facing Low-Income People When Navigating the Financial World* (Toronto: Houselink Community Homes, 2014), http://www.houselink.on.ca/wp-content/uploads/Welcome-to-the-Financial-Mainstream.pdf.

44 Walks, "Mapping."

45 Graham, Hawes, and Ley, *Metro Vancouver's Debtscape.*

46 Ibid., and Walks, "Mapping."

47 Graham, Hawes, and Ley, *Metro Vancouver's Debtscape*, 40.

48 Joe Hermer and Janet Mosher, *Disorderly People: Law and the Politics of Exclusion in Ontario* (Halifax: Fernwood, 2002), 17.

49 Mayer, "First World Urban Activism," 5–19.

50 Mayer and Kunkel, "Introduction."

Toronto's Unsecure(d) Urban Debtscape

Alan Walks

IN DECEMBER 2014, the *Toronto Star* covered a story of a Toronto woman who took out a $10,000 consolidation loan with CitiFinancial, a branch of the large US bank Citigroup. Despite seven years of payments, at virtually 30 percent interest and with additional fees tacked on (including $2,600 in insurance premiums added to the principal), her debt grew to $25,000.[1]

While the details in this case remain unclear, this situation is not uncommon, particularly among lower-income households. Indeed, debt in Canada is currently at its highest level ever recorded and alongside rising income inequality has emerged as one of Canada's most salient issues. Between the early 1980s and the mid-2000s, the ratio of household debt to disposable (after-tax) income more than doubled. After the global financial crisis, household debt continued to rise and by mid-2016 sat at roughly 170 percent of disposable income, more than two-and-a-half times its level during the early 1980s (Figure 1). Some commentators worry that many Canadians will never be able to pay back their debt and that rising bankruptcies will endanger the Canadian economy going forward.[2]

Households in Toronto are even more indebted than the Canadian average. In 2007, before the financial crisis (and the first year for which total household debt is available at this scale), households had total debt representing, on average, approximately 186 percent of their annual disposable income. By 2012, their debt had grown to 193 percent, a 5.5 percent increase in just five years.[3] There are a number of reasons for the rise in household debt, including declining interest rates (which make borrowing cheaper and allow households to easily roll over and renegotiate their debt instead of paying it), stagnant wages among lower-income

FIGURE 1 · **Total household debt as a percent of disposable income, Canada, 1984–2016**

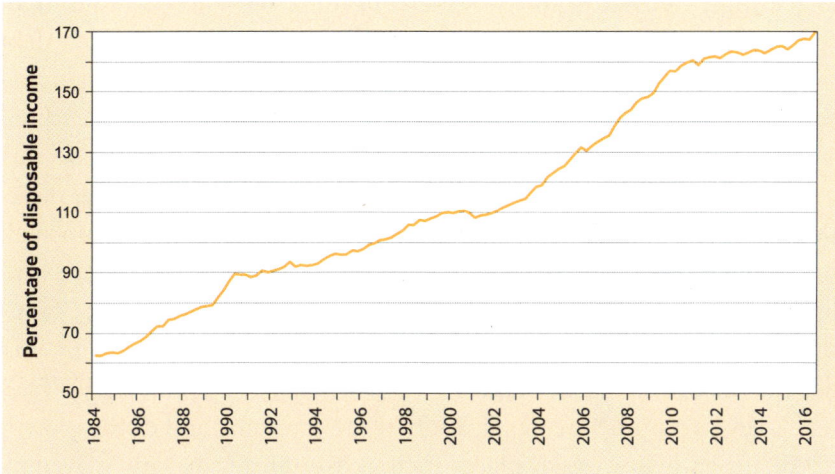

SOURCE: Created by the author, using data from Statistics Canada, CANSIM II Tables 3780051 and 3800019 (1984–89), and CANSIM II Table 3780123 (1990–2016).

households, federal housing policies that have encouraged Canadian mortgage lenders to increase the amount they lend for housing purchases, and an influx of new lenders (many only active online), which has made borrowing easier than ever before.[4] Here, I examine the sociospatial distribution of unsecured debt in Toronto, a component of what I call Toronto's urban debtscape.

THE URBAN DEBTSCAPE

Debt is not distributed evenly, either among social groups or across space. Indeed, in urban Canada, the burden of household debt is higher among lower-income households, immigrants, and single-parent households and, by extension, lower-income neighbourhoods, immigrant-reception neighbourhoods, and neighbourhoods concentrated with families with small children.[5]

The urban debtscape as a concept does not only relate to the distribution of debt and its implications across urban space for rising debt burdens, local housing markets, and vulnerabilities to bankruptcy and foreclosure; it also encompasses the processes and structural conditions that produce debt, including the behaviour of financial institutions, hedge funds, private investors, and household borrowers. Many of these processes fall under the rubric of "financialization," by which profit is increasingly derived from investment in financial markets rather than commodity production.[6] The structural conditions under which debt is produced, consumed and distributed involve changes within national and global economies, including shifts away from investment in capital goods production and towards nationally traded consumption sectors such as housing. They also involve shifts in the type, origin, and direction of financial flows and in each city's strategic location within such financial networks. The urban debtscape is produced in accordance with uneven national and local policies on finance and debt relations, housing markets, and investments; thus, it is partially characterized by, and dependent upon, housing policy.

A key part of the financialization of the economy, and of the growth of the urban debtscape in places such as Toronto, is the role played by political ideologies, attitudes towards urban development and property ownership, and citizen subjectivities with regard to borrowing and saving. How debt (and the responsibilities and liabilities it entails) is internalized and how it affects future behaviour both factor into the evolution of urban debtscapes. The social, cultural, and political (not to mention economic) implications of rising indebtedness are complex and unpredictable and are distributed across space in variegated ways, producing distinctive patterns in each city.[7]

THE NEIGHBOURHOOD DISTRIBUTION OF UNSECURED DEBT

Household debt takes many forms. The vast majority of outstanding household debt in Canadian cities is made up of debt secured by housing in some way, either through mortgages or through home equity lines of credit. In the city of Toronto, these forms of debt represented 84 percent of households' liabilities in 2012. Because these forms of debt are only available to homeowners, they are higher in places with concentrated home ownership. They typically (but not always) entail lower interest rates than other forms of debt, and home owners (or lenders, in the case of foreclosure) can usually sell if they need to pay down their debt. However, when homeowners in places such as the United States, Ireland, and Spain were unable to sell their homes during and after the financial crisis, housing prices dropped precipitously, and many homeowners found themselves "under water" – owing more on their loan than the properties on which they had been secured were worth.[8]

The category of debt not secured by housing comprises a vast grab bag of loan types: credit card debt, unsecured lines of credit, automobile loans and other bank loans, student loans, instalment loans (including for furniture purchases), consolidation loans (including those offered through online lenders), and payday loans. Unsecured forms of debt can be problematic for households, particularly those with low income, because they often have no assets to sell to pay off the debt if the need arises. Even when the loans are used to purchase an asset, such as an automobile or a piece of furniture, the value of these assets usually diminishes rapidly on purchase, so selling them will not produce enough revenue to pay off the loan. Even among home-owning households, the ability to sell a house or to withdraw equity from a house (through mortgage renegotiation or a home-equity line of credit) in order to pay down unsecured debt is diminished for low-income households, who typically face more difficulties in credit markets and who often own more marginal forms of housing. The diversity of loan types within the category of unsecured debt corresponds with the diversity of business models among lenders, and predatory forms of lending are more common among unsecured forms of debt than with debt secured by housing.

The mapping of unsecured forms of debt can thus provide an important window on the distribution of debt-based vulnerabilities across the city. An important indicator is the debt-to-income ratio. Since debt is typically repaid from household income, the debt-to-income ratio gives an indication of the relative burden of debt from household to household (and, when averaged out among local households, from neighbourhood to neighbourhood) after controlling for their income. Disposable income is the income that remains after taxes are paid; it is the actual income available to households for debt repayment. Another indicator of debt burden is the proportion of monthly income going to monthly debt payments; unfortunately, this information is not readily available. Nor are the interest rates paid by individual households on each form of debt. The debt-to-income ratio remains the best indicator available.

Figure 2 shows the neighbourhood (census tract) distribution of average household debt-to-income ratios of debt not secured by housing within the City of Toronto at the end of 2012.[9] Clearly, higher debt ratios are not randomly distributed but instead follow a distinct pattern.

Debt-to-income ratios are higher in working-class neighbourhoods and follow what has traditionally been called the "U" of poverty, named after the U-shaped grouping of neighbourhoods clustered along the Grand Trunk Railway line (where much industrial land, and, in turn, working-class communities, are located), from Union Station to the northeast and northwest of the city. Neighbourhoods with high debt-to-income ratios are found in the West End, including those in Parkdale, the Junction area, Weston, and in the west end of North York up Jane Street. In the East End, such neighbourhoods include those in East York, the Upper Beaches area, and parts of southern and central Scarborough. High levels of unsecured debt are as likely to be found in Toronto's postwar inner suburbs (Scarborough, North York, and Etobicoke) as in the prewar inner city (York, East York, and Old Toronto).

Toronto's neighbourhood debt-to-income ratios correlate with a number of sociodemographic variables at the census tract level. The Pearson correlation coefficient (r) shows the correlation between the census tract averages for each variable, and the distribution of unsecured debt-to-income ratios, without controlling for any other variables or effects.[10] Using this measure, the strongest correlations are with variables related to class position. Neighbourhoods containing households with higher income levels ($r=-0.709$) and dwelling values ($r=-0.708$) are likely to have lower average debt-to-income ratios, while neighbourhoods with greater concentrations of persons with disposable incomes under the low-income cut-off (often used as a poverty line) ($r=0.522$) reveal higher debt burdens. Occupational concentrations also show up as significant, with neighbourhoods that contain higher proportions of those employed in managerial and administrative jobs showing much less unsecured debt ($r=-0.690$), while areas with manufacturing workers show moderately more ($r=0.379$). There is also a moderate correlation with housing tenure, with higher levels of unsecured debt found in neighbourhoods with more tenants ($r=0.315$), which is understandable given that homeowners can often consolidate unsecured debts into their mortgages. Higher unsecured debt levels are also more often associated with neighbourhoods where the greater proportion of adults have only a high school education ($r=0.554$), or less than high school ($r=0.503$), and less often when they hold a university degree ($r=-0.611$). As well, the higher the neighbourhood proportion of immigrants ($r=0.427$) and visible minorities ($r=0.390$), the higher the average unsecured debt-to-income

FIGURE 2 · Average outstanding unsecured household debt (as percent of disposable income), by census tract, City of Toronto, 2012

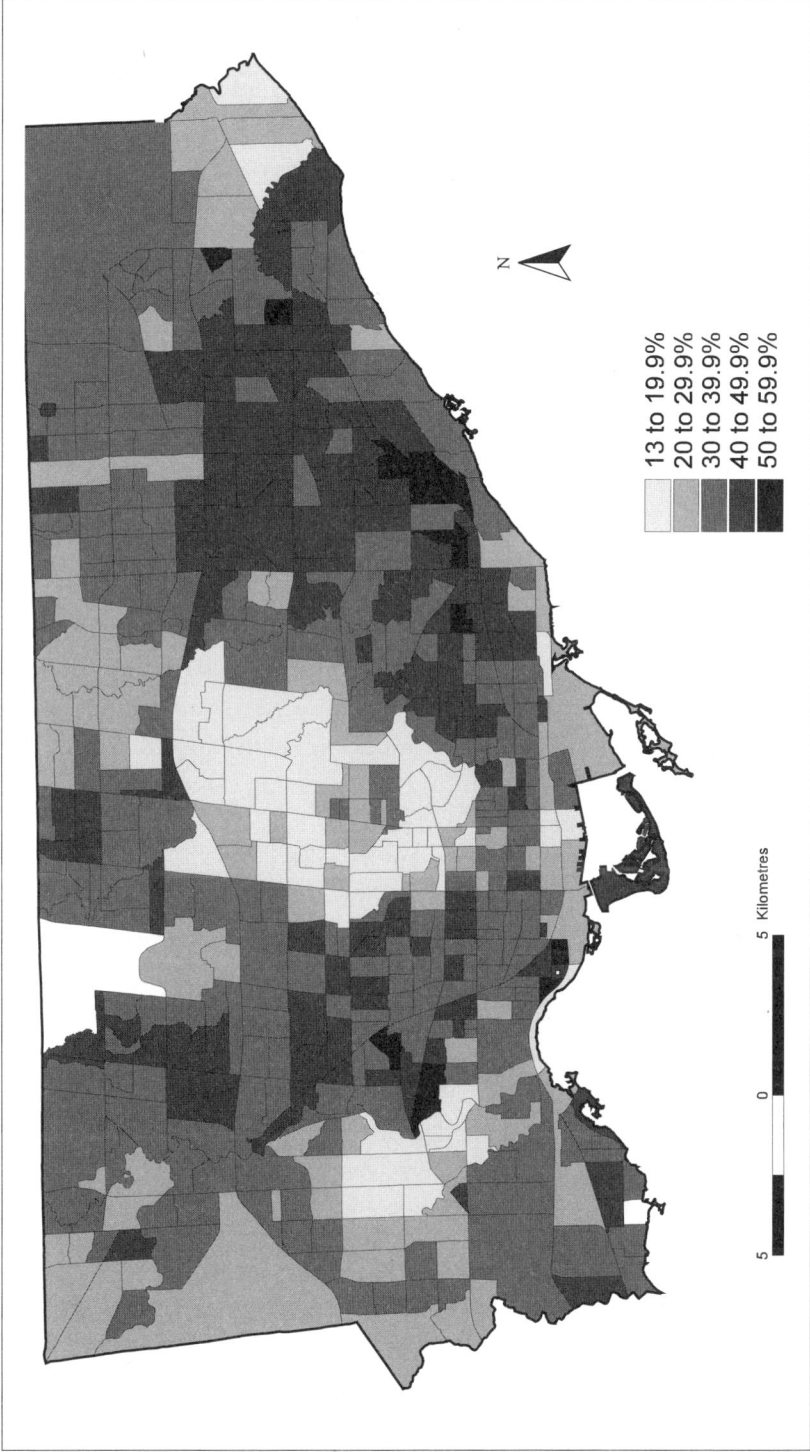

13 to 19.9%
20 to 29.9%
30 to 39.9%
40 to 49.9%
50 to 59.9%

0   5 Kilometres

source: Created by the author, using data from Environics Analytics custom-aggregated to census tracts.

burden, although these correlations are not as strong as those based on class position, and there is virtually no correlation with concentrations of Chinese and South Asian people. Unsecured debt can also be found in neighbourhoods where there are families with young children at home ($r = 0.245$), although this is a weak correlation. Despite these other effects, in general the most consistent pattern is that higher burdens of unsecured debt are found in working-class neighbourhoods.

THE SPATIAL DISTRIBUTION OF DEBT AND THE GLOBAL FINANCIAL CRISIS

The global financial crisis and the federal government's responses to it brought about a shift in the composition of household debt in Canada. Not only did the Bank of Canada bring down its target interest rate considerably (from 3.0 percent to 0.25 percent), the government also encouraged banks to issue mortgages with high loan-to-value ratios and to package them into mortgage-backed securities.[11] It became easier to access mortgages, and those who owned or purchased property could often take out lower-interest mortgages to consolidate outstanding unsecured debts. Those who could not afford home ownership could not take advantage of this situation, leading to increasing class-based discrepancies in access to different forms of credit.

The five-year period between 2007 and 2012 represented a watershed in debt-based class relations in Canada and, consequently, in the evolution of Canada's urban debtscape. In 2007, the year before the financial crisis exploded into full view following the collapse of US-based global investment bank Lehman Brothers, household debt not secured by housing sat at 32.9 percent of household disposable income in the city of Toronto. By 2012, it had fallen to 30.6 percent. As a proportion of total outstanding household debt, debt not secured by housing fell even more: although unsecured debts were falling, mortgages and secured lines of credit were growing. Proportionately, in 2012, unsecured household debt represented about 15.6 percent of total outstanding debt, down from 17.7 percent in 2007.

However, debt burdens did not fall everywhere and in fact rose considerably in a number of places, as Figure 3 shows. The debt-to-income ratio in the downtown area declined substantially, particularly in those neighbourhoods where new condominium housing had been

FIGURE 3 · Change in average outstanding unsecured household debt (as percent of disposable income), Toronto, 2007–12

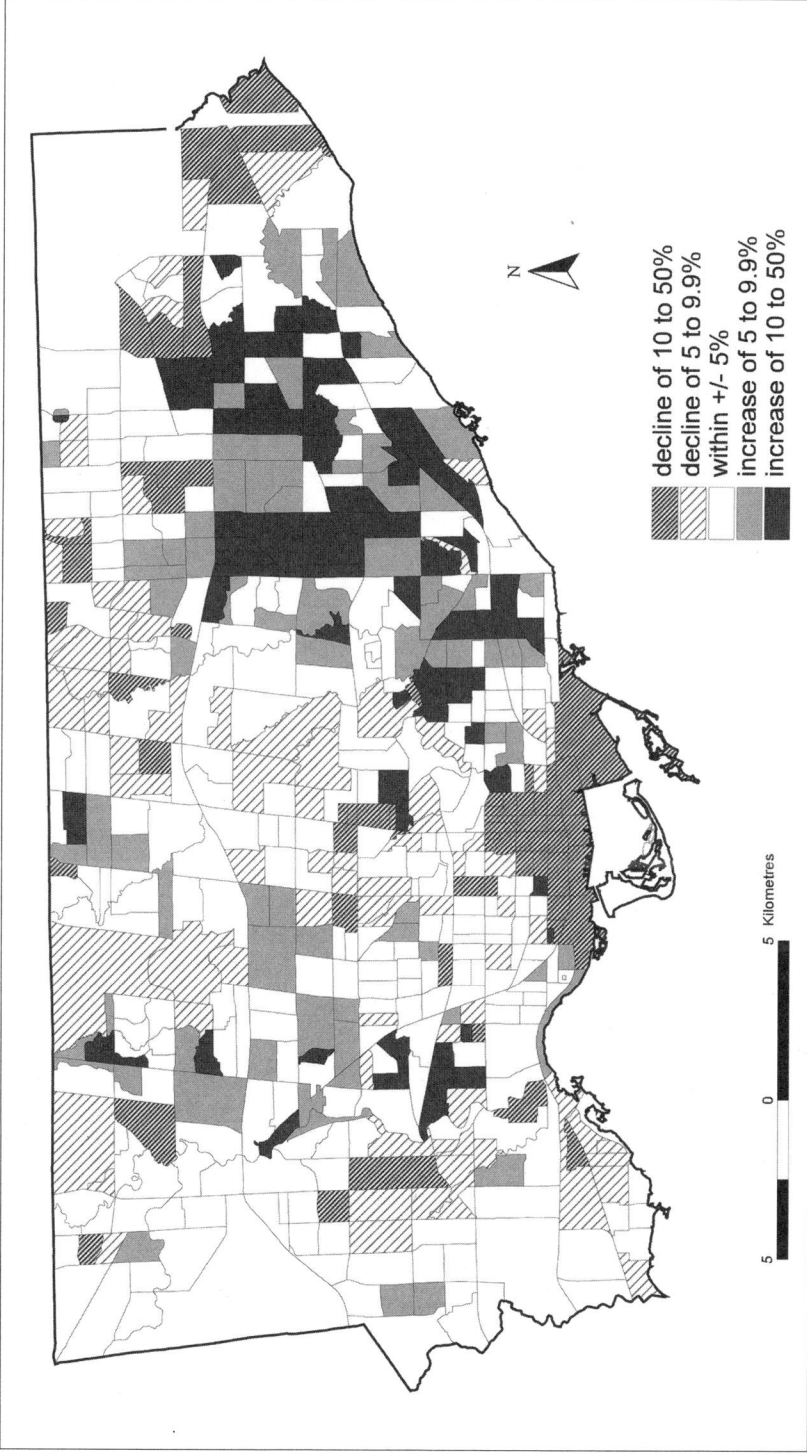

decline of 10 to 50%
decline of 5 to 9.9%
within +/- 5%
increase of 5 to 9.9%
increase of 10 to 50%

SOURCE: Created by the author, using data from Environics Analytics custom-aggregated to census tracts.

built since 2000.[12] Indeed, the presence of condominiums and high-rise residential towers (greater than five storeys high) are two of the strongest predictors of declining unsecured debt levels over the 2007–12 period. Rising property values, partially due to federal government policies, helped households in these areas reduce their unsecured debts, perhaps by consolidating them into their mortgage debt.[13] Furthermore, as gentrification through condominium development enveloped the downtown core, higher-income residents with lower levels of debt concentrated there. Neighbourhoods that witnessed a decline in unsecured debt levels of at least 5 percentage points had average household incomes roughly 25 percent higher than other neighbourhoods. Although some of the neighbourhoods undergoing gentrification – the Beaches, Riverdale, parts of Bloor West Village, and others – do reveal rising debt-to-income ratios for unsecured debt, the majority of neighbourhoods suffering increasing debt burdens are located in the older postwar inner suburbs, such as the former city of Scarborough, old Weston, and the Jane-Finch Corridor.

Rising debt burdens reveal important sociospatial patterns. Notably, neighbourhoods with increasing debt levels are more likely to have lower incomes: on average, in 2012, incomes were 30 percent lower in neighbourhoods where debt had increased by 5 percentage points or more than in other neighbourhoods. Furthermore, those neighbourhoods that saw unsecured debt rise 5 percent or more also had higher poverty rates (roughly 3 percent higher, on average), higher concentrations of manufacturing workers (8.1 percent versus 6.6 percent) and visible minorities (48.6 versus 42.9 percent), and lower dwelling values (22 percent lower) than other neighbourhoods.[14] Other variables – including immigration status, age, and housing tenure – showed no statistically significant effects, suggesting that rising debt burdens are mostly felt in neighbourhoods with poorer, working-class, and racialized communities.

· · · · ·

Toronto's urban debtscape is in a process of evolution. Studies of inequality rarely take the distribution of debt and wealth into account, largely because data have not been widely available and debt payments come out of after-tax income. But it is critical to incorporate debt and wealth into perspectives on inequality, including those of other social groups and those pertaining to the neighbourhood segregation of income. Indeed, with rising levels of household debt, debt itself is becoming an indicator of a new social cleavage, with high debt levels carrying implicit vulnerabilities, but also imparting new forms of possible advantage or disadvantage, depending on a household's access to various different forms of credit. Debt not secured by housing typically involves higher interest rates, leaving households that may not have sufficient assets vulnerable. A homeowner who loses a job can usually sell the house and rent or move to a less expensive one. Poorer households are more likely to be tenants, may not own an automobile, and may be forced to choose between making their debt payments or paying for groceries, rent, or heating. Those with high levels of unsecured debt are more vulnerable to predatory lenders (which may even include mainstream banks), who may offer new loans at usurious interest rates and on terms that effectively prevent any escape from debt. In these circumstances, debtors may feel more compelled to take unsafe and poorly paid work. Debt relations are therefore an important contributor to patterns of inequality and poverty, and the production and evolution of the urban debtscape.

NOTES

1 Dana Flavelle, "Debt Struggles Spark Concerns with Activists," *Toronto Star*, December 6, 2014.
2 See, for example, Hilliard Macbeth, *When the Bubble Bursts: Surviving the Canadian Real Estate Crash* (Toronto: Dundurn Press, 2015).
3 Calculated by the author from Environics Analytics' "Wealthscapes" data sets for each year.
4 Alan Walks, "Canada's Housing Bubble Story: Mortgage Securitization, the State, and the Global Financial Crisis," *International Journal of Urban and Regional Research* 38, 1 (2014): 256–84; and Robb Engen, "Online Lenders Target Those Needing Instant Money," *Toronto Star*, July 12, 2016.
5 Matt Hurst, *Debt and Family Type in Canada* (Ottawa: Statistics Canada, 2011); component of Catalogue no. 11-008-X, April 21, 2011; and Alan Walks, "Mapping the Urban Debtscape: The Geography of Household Debt in Canadian Cities," *Urban Geography* 34, 2 (2013): 153–87.
6 Greta Krippner, "Financialization of the American Economy," *Socio-economic Review* 3, 2 (2005): 173–208.
7 Alan Walks, "Mapping the Urban Debtscape," and Walks, "Homeownership, Asset-Based Welfare and the Neighbourhood Segregation of Wealth," *Housing Studies* 31, 7 (2016): 755–84.

8   Alan Walks and Dylan Simone, "Unequal and Volatile Urban Housing Markets," in *Urbanization in a Global Context*, ed. Alison Bain and Linda Peake (Toronto: Oxford University Press, 2017), 190–208.

9   Census tracts are spatial units created by Statistics Canada as a proxy for neighbourhoods. Each tract contains between four thousand and eight thousand people.

10 Pearson correlation coefficients vary in strength between 0 (no correlation) and 1 (complete correspondence) and can be positive or negative. The results are thus a measure of *spatial* correlation.

11 Walks, "Canada's Housing Bubble Story."

12 Gillad Rosen and Alan Walks, "Castles in Toronto's Sky: Condoism as Urban Transformation," *Journal of Urban Affairs* 85 (1) (2015): 39–66.

13 Ibid., and Alan Walks and Brian Clifford, "The Political Economy of Mortgage Securitization and the Neoliberalization of Housing Policy in Canada," *Environment and Planning A* 47, 8 (2015): 1624–42.

14 Data for 2012 household income come from the same custom data set as the debt data (Environics Analytics). The other variables derive from the 2006 Census of Canada. While it would have been best to have more up-to-date census data, the Canadian federal government under former prime minister Steven Harper cancelled the long-form census in 2011, leaving the 2006 census as the most recent reliable and relevant census data at the neighbourhood scale to the 2012 debt data, at the time of writing.

Automating Social Inequality

Krystle Maki

IN THE MID-1990S, the Province of Ontario enacted new social assistance legislation. It rebranded the welfare program as "Ontario Works," introduced work-for-welfare requirements, initiated anti-fraud measures and rigorous surveillance of recipients, and imposed zero-tolerance policies and lifetime bans on those convicted of welfare fraud. The cost overhaul produced a punitive welfare system aimed at criminalizing poverty and discouraging the poor from seeking state support.[1]

Motivated by the Ontario government's Business Transformation Project and partnership with Accenture, a private IT corporation, the official marketing of these policy reforms shifted the discourse on public social services and the welfare of the poor towards business efficiency, individualism, and neoliberal public management.[2] Feminist criminologists and social legal scholars have demonstrated that the dismantling of the welfare state and the increased surveillance and criminalization of marginalized groups are strategies to increase and justify regulation over certain "othered" groups, particularly racialized women and single mothers.[3] Shifting legal discourse "from welfare fraud to welfare as fraud" condemns social assistance and its recipients as criminal thus "linking poverty, welfare and crime."[4]

As in the United States and other jurisdictions, Ontario's neoliberal social assistance reforms have been enabled materially by a succession of automated systems, purchased and implemented through contracts with private corporations active in the new global market of neoliberal welfare state restructuring. The costs of these new surveillance technologies are significant: by 2005, Accenture's costs exceeded $377 million, $197 million over budget.[5] Material and discursive shifts in social assistance are

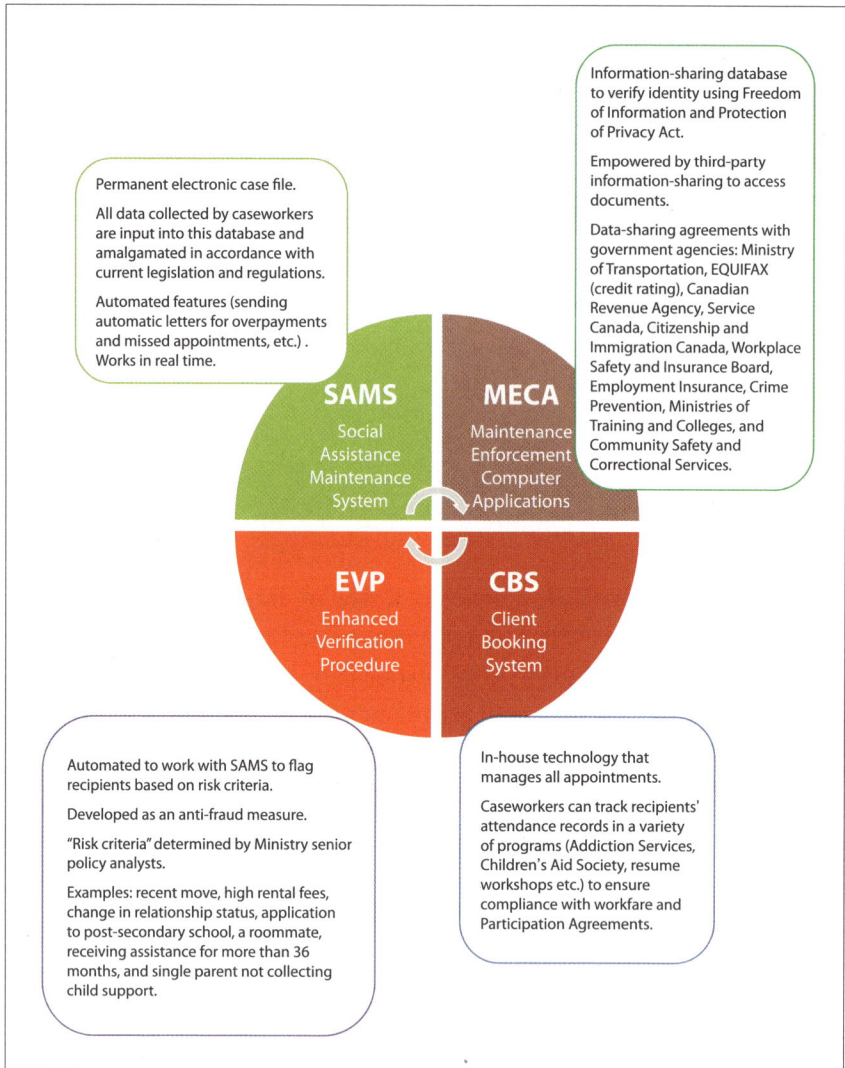

Permanent electronic case file.

All data collected by caseworkers are input into this database and amalgamated in accordance with current legislation and regulations.

Automated features (sending automatic letters for overpayments and missed appointments, etc.) . Works in real time.

Information-sharing database to verify identity using Freedom of Information and Protection of Privacy Act.

Empowered by third-party information-sharing to access documents.

Data-sharing agreements with government agencies: Ministry of Transportation, EQUIFAX (credit rating), Canadian Revenue Agency, Service Canada, Citizenship and Immigration Canada, Workplace Safety and Insurance Board, Employment Insurance, Crime Prevention, Ministries of Training and Colleges, and Community Safety and Correctional Services.

SAMS
Social Assistance Maintenance System

MECA
Maintenance Enforcement Computer Applications

EVP
Enhanced Verification Procedure

CBS
Client Booking System

Automated to work with SAMS to flag recipients based on risk criteria.

Developed as an anti-fraud measure.

"Risk criteria" determined by Ministry senior policy analysts.

Examples: recent move, high rental fees, change in relationship status, application to post-secondary school, a roommate, receiving assistance for more than 36 months, and single parent not collecting child support.

In-house technology that manages all appointments.

Caseworkers can track recipients' attendance records in a variety of programs (Addiction Services, Children's Aid Society, resume workshops etc.) to ensure compliance with workfare and Participation Agreements.

Ontario Works surveillance matrix operating system

made visible by the matrix of digital systems that rank applicants' and recipients' information against risk criteria to verify identity and track fraudulent activity.

In the Ontario Works surveillance matrix, applicants are required to sign over their privacy rights and submit extensive documentation, including birth certificates, divorce papers, immigration papers, and documents related to student loans, housing, bank accounts, addiction treatment, and paternity tests.[6] This personal information is uploaded to Ontario Work's databases to amalgamate, data mine, and assess applications using risk criteria and eligibility verification procedures. This personal information is then shared with dozens of government departments and institutions through third-party agreements. These automated features are powerful surveillance tools that combine technology with caseworkers' assessments to automate inclusion and exclusion for social assistance.

The welfare application process begins with the Social Assistance Management System (SAMS), designed to collect and analyze a range of documents, information from third parties, records of workfare participation and addictions treatments, and any additional documentation requested by caseworkers. Ultimately, SAMS is a permanent electronic version of a recipient's file that hundreds of government employees can access, drastically widening the scope of which government bodies can access information about those on social assistance.

Once information is recorded in SAMS, it is processed by the maintenance enforcement computer application (MECA), which is empowered through third-party verifications to do the following: aid law enforcement investigations, locate individuals for recovery of overpayments, authenticate credit ratings, review student loans and status, confirm probation conditions, and determine initial and ongoing eligibility for social assistance by verifying the recipient's personal information. The recipient's information is then analyzed by the eligibility verification procedure (EVP), a database that ranks recipients' information and flags files on predetermined risk criteria developed by policy analysts at the Ministry of Community and Social Services with reference to the current economic climate and Ontario Works quotas tied to provincial funding.[7] These three systems and applications – SAMS, MECA, and the EVP – organize data to uncover hypothetical fraudulent activities, which can result in a suspension or termination of benefits, investigations by the eligibility review officers (the fraud unit), or, in some cases, criminal charges, thereby intensifying the criminalization of poverty.

The Ontario Works recipients I interviewed revealed that they were regularly red-flagged, investigated, and called in for reviews, acts that generated a climate of stress, suspicion, surveillance, control, and fear for those living on the economic margins of society, fear that their benefits

Information requests	Documentation to verify
· How many people are in your family, and what are their ages?	· birth certificates or passports
	· health cards
	· Indian status cards
· Immigration status?	· immigration documents
· What type of housing do you live in?	· rent receipts and leases or tenancy agreements
	· mortgage agreements and statements
· What education do you have?	· other bills related to housing costs, such as bills for hydro, water, gas, property taxes, and home or apartment insurance
· Are you working now? If not, what were your past jobs?	
	· bank statements or bank records
· Do you have income of any kind, including money from a job, support payments, or benefits such as Canada Pension Plan or Employment Insurance?	· evidence of income from any source, including employment, support payments, workers' compensation, and payments from tenants, roomers, and boarders
	· income tax notices, called Notice of Assessment from Canada Revenue Agency
· Do you have assets, including money in bank accounts, registered retirement savings plans (RRSPs), registered education savings plans, guaranteed investment certificates, insurance policies, or vehicles?	· information about assets, including RRSP statements, car ownership papers, life insurance policies, and bonds
	· pay stubs, records of employment, and letter of termination
	· documents showing school attendance
	· credit card bills or advances
· Do you have debts, including money owed on credit cards, to the bank, or to other people?	· student loans, such as loans from the Ontario Student Assistance Program
	· other loans, even those from relatives or friends

SOURCE: "Ontario Works Directive 2.1: Application Process." Ministry of Children, Community and Social Services. https://www.mcss.gov.on.ca/en/mcss/programs/social/directives/ow/2_1_OW_Directives.aspx (accessed December 12, 2019).

could be suspended pending an investigation. For instance, Diane, a single mother, underwent intensive questioning from her caseworker because her file had triggered risk criteria. Diane's file was flagged because her rent was considered higher than Ontario Work's shelter allowance (which was $376 for a single parent in 2015). Her caseworker questioned how she could afford to live on such a small budget, implying that she was getting help from somewhere else or defrauding the government. Diane responded to her caseworker, "Most days I go without. I starve, okay. I exhaust food banks."[8] Often, recipients were unaware they were doing anything wrong. They felt violated and feared for their economic survival. While these systems are allegedly in place as anti-fraud measures and to ensure that the system works efficiently to weed out the ineligible, there are life-altering consequences for those who rely on benefits for their daily needs.

In the past two decades, surveillance of Ontario Works recipients has relied on technologies to determine ongoing eligibility for benefits and to reduce caseloads. Yet former regulatory techniques have also persisted. The public is encouraged to anonymously report suspicious activities to the Welfare Fraud Hotline, which scholars have discovered targets single mothers.[9] All calls are followed up with an investigation by eligibility review officers in the fraud unit. The fraud hotline insinuates linkages between poverty, social assistance, and criminality as the public is invited into the surveillance gaze to catch welfare fraudsters in the form of community surveillance.[10]

Welfare policies have always been concerned with moral regulation and, aside from any benevolent aims, have primarily served as methods of social control.[11] Technological advancements in welfare surveillance have created more openings for investigation, termination of benefits, and even criminalization. But the rationale behind these advancements represents a continuity with, not a break from, previous welfare regimes, which, since the uneven introduction of the British Poor Laws and workhouses in the colonies in the 1830s, have aimed at fiscal restraint by providing the bare minimum of benefits to compel recipients into low-paid employment. Even during the height of the Keynesian welfare state (1940–73), recipients, especially single mothers and racialized recipients, faced state surveillance and had to prove their deservedness.[12] The automation of surveillance, beginning in the 1990s, was prompted by a neoliberal rollback of the social safety net and a return to the type of welfare fraud hysteria that had not been seen in Canada since the Great Depression.

Welfare policies and state surveillance are experienced differently, depending on a person's gender, race, and relationship status; women, women of colour, and single mothers are disproportionately impacted by surveillance because of enforced paternity testing, spouse-in-the-house rules that monitor their sexuality, child support requirements that oblige them to interact with ex-partners, and the lack of affordable and accessible child care necessary to comply with workfare duties.[13]

Ontario Work's workfare model, and the technologies it relies upon, have normalized the monitoring, surveillance, and regulation of welfare recipients, denying them rights to privacy and concealing the systemic factors that shape inequality. Welfare surveillance is purposeful: it structures the political economy of how social assistance functions and shapes moral discourses about recipients. In many ways, welfare surveillance has automated social inequality in Ontario and institutionalized monitoring and regulation of the poor.

NOTES

1   See Jannie Abell, "Structural Adjustment and the New Poor Laws: Gender, Poverty and Violence and Canada's International Commitments," *Canadian Feminist Alliance for International Action, FAFIA Think Tank Paper* No 3 (2001): 5–38; Maureen Baker, "The Restructuring of the Canadian Welfare State: Ideology and Policy," SPRC Discussion Paper 77, Social Policy and Research Centre, Sydney, 1997; Gillian Balfour and Elizabeth Comack, *The Power to Criminalize: Violence, Inequality and Law* (Halifax: Fernwood Publishing, 2004); Janine Brodie, *Politics on the Margins: Restructuring and the Canadian Women's Movement* (Halifax: Fernwood, 1996); Lea Caragata, "Neoconservative Realities: The Social and Economic Marginalization of Canadian Women," *International Sociology* 18, 3 (2003): 559–80; Dorothy Chunn and Shelley Gavigan, "Welfare Law, Welfare Fraud and the 'Never Deserving' Poor," *Social Legal Studies* 13, 2 (2004): 219–43; Janine Fitzgerald, "The Disciplinary Apparatus of Welfare Reform," *Monthly Review*, November 2004, https://monthlyreview.org/2004/11/01/the-disciplinary-apparatus -of-welfare-reform/; Kiran Mirchandani and Wendy Chan, *Criminalizing Race, Criminalizing Poverty: Welfare Enforcement in Canada* (Nova Scotia: Fernwood, 2007); Janet Mosher, Patricia Evans, Margaret Little, Eileen Morrow, Jo-Anne Boulding, and Nancy VanderPlats, *Walking on Eggshells: Abused Women's Experiences of Ontario's Welfare System*, final report of research

findings from the Women and Abuse Welfare Research Project (Toronto: York University, 2004), http://www.yorku.ca/yorkweb/special/Welfare_Report _walking_on_eggshells_final_report.pdf; and Sherri Torjman, *Workfare: A Poor Law*, report for Caledon Institute of Social Policy, February 1996, https://maytree.com/wp-content/uploads/10ENG-1.pdf.

2 Krystle Maki, "Automating Social Inequality: How Single Mothers and Caseworkers Navigate the Neoliberal Surveillance of 'Ontario Works,' 1995–2015" (PhD diss., Queen's University, 2015).

3 See Abell, "Structural Adjustments"; Balfour and Comack, *The Power to Criminalize*, and *Criminalizing Women: Gender and (In)justice in Neoliberal Times* (Halifax: Fernwood, 2006); Gillian Balfour, "Introduction: Regulating Women and Girls," in *Criminalizing Women*, 154–73; Helen Boritch, *Fallen Women: Female Crime and Criminal Justice in Canada* (Scarborough: International Thomas Publishing, 1997); Chunn and Gavigan, "Welfare Law"; Kelly Hannah-Moffat, "Prisons That Empower: Neo-liberal Governance in Canadian Women's Prisons," *British Journal of Criminology* 40, 3 (2000): 510–31; Margaret Little, "The Leaner, Meaner Welfare Machine: The Harris Government's Ideological and Material Attack on Single Mothers," in *Making Normal: Social Regulation in Canada*, ed. Deborah Brock (Toronto: University of Toronto Press, 2003), 235–58; Jill McCorkel, "Criminally Dependent? Gender, Punishment, and the Rhetoric of Welfare Reform," *Social Politics: International Studies in Gender, State and Society* 11, 3 (2004): 386–410; Mirchandani and Chan, *Criminalizing Race*; Janet Mosher and Joe Hermer, "Welfare Fraud: The Constitution of Social Assistance as Crime," *Commissioned Reports and Studies*, 2005, Paper 161, https://digitalcommons. osgoode.yorku.ca/reports/161; and Kim Pate and Debbie Kilroy, "Developing International Norms and Standards to Meet the Needs of Criminalized and Imprisoned Women," paper prepared for the Canadian Association of Elizabeth Fry Societies, 2005, http://www.caefs.ca/resources/issues-and -position-papers/united-nations/.

4 Chunn and Gavigan, "Welfare Law," 219.

5 Todd Gabel, Jason Clemens, and Sylvia LeRoy, "Welfare Reform in Ontario: A Report Card," Fraser Institute Digital Publication, September 2004, 22, https://www.fraserinstitute.org/studies/welfare-reform-ontario-report -card; Richard Girard, "Accenture Profile," *Polaris Institute*, 2003, http://www. polarisinstitute.org/files/Accenture.pdf (accessed February 8, 2016); and Office of the Auditor General of Ontario, annual report, 2009, Chapter 3.11, Ontario Works Program, http://www.auditor.on.ca/en/reports_2009_ en.htm (accessed February 3, 2015).

6 *Ontario Works Act, 1997*, SO 1997, c 25, Sched A, O Reg 134/98: General, https://www.ontario.ca/laws/regulation/980134.

7 Maki, "Automating Social Inequality."

8 Diane [a pseudonym], interview with author, Peterborough, December 6, 2011.

9 Little, "The Leaner, Meaner Welfare Machine," and Mosher et al., *Walking on Eggshells.*

10 Maki, "Automating Social Inequality."

11 John Gilliom, *Overseers of the Poor: Surveillance, Resistance, and the Limits of Privacy* (Chicago: University of Chicago Press, 2001); Margaret Little, "'Manhunts and Bingo Blabs': The Moral Regulation of Ontario Single Mothers," *Canadian Journal of Sociology* 19, 2 (1994): 233–47; and Francis Piven and Richard Cloward, *Regulating the Poor: The Functions of Public Welfare* (New York: Pantheon, 1971).

12 Margaret Little, *No Car, No Radio, No Liquor Permit: The Moral Regulation of Single Mothers in Ontario, 1920–1997* (Toronto: Oxford University Press, 1998); Mirchandani and Chan, *Criminalizing Race*; and Brian Palmer and Gaetan Heroux, *Toronto's Poor: A Rebellious History* (Toronto: Between the Lines, 2016).

13 Caragata, "Neoconservative Realities"; Margaret Little, "A Litmus Test for Democracy: The Impact of Ontario Welfare Changes on Single Mothers," *Studies in Political Economy* 66 (2001): 9–35; Maki, "Automating Social Inequality"; and Mirichandani and Chan, *Criminalizing Race*.

ACORN's Campaign for Affordable Access

Judy Duncan, ACORN

COUNTRIES AROUND THE world increasingly recognize access to the internet as an essential tool for participation in modern democratic society. Access to reliable high-speed internet has become an important means of participating in economic, social, and civic life.

In Canada, though, internet access is deeply unequal. According to Statistics Canada, 42 percent of households in the lowest income quartile – those who earn $30,000 or less – do not have home internet access. In contrast, nearly all households in the highest income quartile have internet access at home – a mere 2 percent do not.[1]

In Canada's market-dominated telecommunications system, competition among internet providers might have once been considered a solution to bridging the digital divide. Unfortunately, this competition has failed to lower the price of home broadband access. The market in Canada is controlled by a small group of firms, and prices remain prohibitively high for many low-income families.

To bridge the digital divide, ACORN members launched a campaign to pressure the federal government and the Canadian Radio-television and Telecommunications Commission to create a mechanism to ensure that home broadband prices are affordable for low-income families. We are advocating for ten-dollars-a-month high-speed internet (fifteen megabits per second or best speed available in the area) and subsidized computers for all individuals and families who are below the low-income measure.

THE RIGHT TO THE INTERNET

The United Nations now considers internet access a human right comparable with freedom of speech. Frank La Rue, The UN special rapporteur on the promotion and protection of the right to freedom of opinion and expression, has stated that the internet is a key means by which individuals exercise their human rights.[2] In 2011, he described the internet as a tool with enormous potential for realizing the right to freedom of expression. A year earlier, a United Nations report identified disparities in access to information and communications technologies – particularly pricing structures that prohibit the poor from accessing the internet – as a key threat to freedom of expression.[3]

A 2010 BBC World Service poll of 27,973 adults in twenty-six countries found that 70 percent of adults regard internet access as a fundamental right. Of Canadians surveyed, 77 percent somewhat agreed or strongly agreed that access to the internet should be a human right.[4] Further, digital access is guaranteed as a legal right in several countries, and many more have publicly announced a commitment to achieving universal broadband access for all citizens. Several countries have adopted laws that either require the state to work to ensure that internet access is broadly available or prevent the state from unreasonably restricting an individual's access to information and the internet. Some countries' laws do both.

In 2013, for example, the German Federal Court of Justice ruled that the internet is a legal right. The court ruled that a man who had lost his connection for two months because of the provider's error was entitled to compensation for the loss of a necessity.[5] In 2009, France's highest court, the Constitutional Council, ruled that new antipiracy legislation – which allowed individuals to be disconnected from the internet without any judicial oversight – infringed on the right to freedom of expression. The council said that freedom of expression includes the right to access online networks, given their importance to participation in democratic life, and that access to the internet is integral to the exercise of other rights and freedoms.[6]

NATIONAL BROADBAND PLANS

Finland was the first country to make high-speed internet a legally enforceable right. Since 2010, every citizen in Finland has had a legal right to broadband internet access at a speed of at least 1 megabyte per second, and it set a goal to achieve a 100-megabytes-per-second connection for everyone by 2015.[7]

Many other countries have since announced national broadband plans. These plans typically set a target for the percentage of citizens who will have access to the internet at a certain connection speed and impose a timeline within which to achieve this goal. Broadband plans may also define the type of technology (e.g., fibre optics) that will be used to achieve the goal, and they may allocate funding for the project.

As part of its ten-year growth strategy, the European Union launched an initiative called Digital Agenda for Europe, which sets a target of bringing internet speeds over 30 megabytes per second to all European citizens by 2020. One of the agenda's action items calls on member states to develop operational national broadband plans.[8] France has had a national digital plan to achieve universal broadband access since 2008.[9] The United Kingdom has set a target of universal broadband of at least 2 megabytes per second for all Britons by 2015. The government has also established a rural community broadband fund, aimed at bringing broadband to the 10 percent of areas that are hardest to reach. The United Kingdom views broadband internet as part of its universal service obligation, and it has said that all Britons should have access to it.[10] Australia's National Broadband Policy set a goal to bring high-speed broadband to all Australians by 2019.[11] South Africa has a draft National Broadband Policy, which is currently in the final stages of consultation, calling for universal broadband access at 5 megabytes per second for all citizens by 2020 and at 100 megabytes per second by 2030.[12]

In the United States, telecommunications companies contribute to the Universal Service Fund based on earned revenue, and money in the fund goes to lower the cost of digital services for low-income households. The US National Broadband Plan calls for up to $15.5 billion to be allocated from the Universal Service Fund over ten years to support increased access to affordable broadband. The Federal Communications Commission is currently testing a pilot program that offers broadband subscribers with low-income a discount on their monthly bill.[13]

THE CAMPAIGN

In fact, Canada is the only G7 nation without a national broadband plan.[14] In June 2015, the Canadian Radio-television and Telecommunications Commission (CRTC) launched a major initiative to "ensure that

Canadians have access to world-class telecommunications services that enable them to participate actively in the digital economy."[15] ACORN Canada has partnered with the Public Interest Advocacy Centre to create the Affordable Access Coalition. Our goal is to develop an official submission and participate in the CRTC's public consultations.

A survey of four hundred ACORN members across Canada found that for 83.5 percent of respondents, the price of home internet is extremely expensive.[16] The most common response (59 percent) was "Extremely high; I can't afford it, but because I need it I take money out of my budget for other items." Comments from respondents revealed that the internet is a particularly vital resource for low-income people, including persons with disabilities, people who are unemployed, and parents. Here are some of the comments:

> As someone with a disability, the internet is essential in accessing disability supports. The Internet is often the only contact point for companies and venues and I need that information to determine if they are accessible and/or can accommodate me. Also, things like paratransit bookings are often only done online.

> It would save me a significant amount of time and energy and allow me to be a competitive player in the process of applying for employment. The time and energy I would save not having to go to the library would allow me to prepare healthy meals and eat at home which helps me to maintain my blood sugars, save money, eat healthier and feel better.

> My children's school WILL NOT OFFER the option of getting important notices on paper and will only communicate updates by email. I have no recourse in this matter and must have an email account I can check regularly. Without internet one cannot meet the expected level of communication and will miss out on many opportunities.

> As my child gets older (she's seven right now), more and more of her special education lessons will require the internet. Once she learns how to read and write by herself she will need [internet]

access to further her learning, and I worry that she'll fall behind because I can't afford anyone's rates.

Working with our members and leaders, we decided to deliver these testimonials in person in a binder to four CRTC offices across the country. Over one hundred ACORN members participated in their delivery. Along with the testimonials, we requested that ten ACORN members be permitted to participate in the ongoing hearings, and the CRTC agreed. No other group or company was given anywhere close to that level of participation. The impact of ACORN's intervention was publicly acknowledged by the CRTC's chair, Jean-Pierre Blais. In an article in Canada's national newspaper, it was said that he "appeared to be personally affected by the presentations so far, noting that he spent the weekend reflecting on some of the 'particularly striking' evidence the CRTC had heard, which included a number of individuals speaking about their own struggles due to having limited or no access to high-speed Internet."[17]

Since the hearing, one of Canada's major digital service providers has announced a product geared towards single, low-income mothers. This is a good step, but its narrow scope outlines the importance of having the CRTC and government mandate an inclusive policy that will reach all low-income households in Canada. In late 2016, the CRTC declared internet a basic service that all Canadians should have access to at home, with the Ministry of Innovation responsible to create a solution to ensure that all people can afford high-speed internet at home.

On June 6, 2018, the Minister of Innovation, Science and Economic Development Navdeep Bains marked a significant step towards realizing ACORN's Internet for All campaign, as he announced a new program offering affordable internet for low-income households. The Connecting Families program targets National Child Benefit recipients and provides 10 megabytes of internet with 100 gigabytes of usage for $10 per month. Around 220,000 households – up to 600,000 people – are expected to benefit, keeping approximately $80 million in the pockets of low-income parents.

ACORN members were pleased that the Federal Government was taking leadership on the issue of digital equity. The Connecting Families program, however, is voluntary, so it will not be offered by all internet service providers. Participating providers include Rogers, Bell, TELUS, Shaw, Vidéotron, SaskTel, and Cogeco. Nova Scotia ACORN members were deeply disappointed to discover that Eastlink opted out of the program, snubbing low-income Nova Scotians. Further, it was not universal, so low-income people without young children did not qualify.

BUILDING PEOPLE'S POWER FOR CHANGE

This campaign serves as a good example of ACORN Canada's organizing model. We spend a large amount of our resources going door to door doing leadership development in low-income neighbourhoods. We engage people on issues they care about, and from there we develop our campaign policies upward. Affordability is a large issue when we talk to people. Our campaigns always include leadership development and planning with low- and moderate-income community members. Leadership Schools, run by members who are leaders, help connect local issues to larger policies. Conversations with community members, surveys, and testimonials from ACORN members help inform our national policy initiatives. These initiatives are approved by our national board, which is made up of ACORN leaders elected by our thousands of members. ACORN's grassroots strategy for encouraging participation in civic processes, like the CRTC hearing, is unique. Engaging organized communities in research and policy development directly and independently is vital to ensuring effective government policy.

Our Fair Banking Campaign offers another example of our bottom-up policy development model. We're fighting for access to fair credit and an antipredatory lending strategy by the federal, provincial, and local governments. Paying for access to the internet is one of the factors driving people into debt. High-speed internet is now a necessity, and people are stretching their budgets to access it. Consumer debt and predatory lending are inextricably linked: for example, people take out payday loans to avoid having their internet cut off. As a result, our Fair Banking Campaign and our Internet for All Campaign are two sides of the same campaign.

This is why ACORN Canada is an organization for low- and moderate-income families focused on building power for change. We'll never arrive at social justice if we focus only on a specific campaign. It will only come when low- and moderate-income people have enough power to change the system!

NOTES

1 Statistics Canada, "Household Access to the Internet at Home, by Household Income Quartile and Geography," Table 22-10-0007-01 (formerly CANSIM 358-0167), https://www150.statcan.gc.ca/t1/tbl1/en/tv.action?pid=2210000701.

2 Frank La Rue, "Report of the Special Rapporteur on the Promotion and Protection of the Right to Freedom of Opinion and Expression," May 16, 2011, A/HRC/17/27, https://undocs.org/en/A/HRC/17/27.

3 United Nations, "Tenth Anniversary Joint Declaration: Ten Key Challenges to Freedom of Expression in the Next Decade," March 25, 2010, A/HRC/14/23/Add.2, https://undocs.org/A/HRC/14/23/Add.2, https://undocs.org/A/HRC/14/23/Add.2.

4 BBC World Service, "Four in Five Regard Internet Access as a Fundamental Right: Global Poll," August 3, 2010, http://news.bbc.co.uk/2/shared/bsp/hi/pdfs/08_03_10_BBC_internet_poll.pdf.

5 DW.com, "Internet Access Declared a Basic Right in Germany," January 27, 2013, https://www.dw.com/en/internet-access-declared-a-basic-right-in-germany/a-16553916; and Reuters, "German Court Rules Internet 'Essential,'" January 24, 2013, http://www.reuters.com/article/2013/01/24/us-germany-internet-idUSBRE90N15H20130124.

6 Conseil Constitutionnel, Decision No. 2009-580DC, June 22, 2009, https://www.conseil-constitutionnel.fr/decision/2009/2009580DC.htm.

7 Saeed Ahmed, "Fast Internet Access Becomes a Legal Right in Finland," CNN, October 15, 2009, http://www.cnn.com/2009/TECH/10/15/finland.internet.rights/; and BBC World Service, "Four in Five."

8 European Community, "Connectivity for a European Gigabit Society," n.d., http://ec.europa.eu/digital-agenda/en/our-goals/pillar-iv-fast-and-ultra-fast-internet-access.

9 Hogan Lovells, "France Launches Ambitious 'Digital Plan,'" Lexology, October 30, 2008, https://www.lexology.com/library/detail.aspx?g=0df83ece-7aaf-4a20-b78c-c42d34eaf23c.

10 United Kingdom, "Broadband Delivery UK: Details of the Plan to Achieve a Transformation in Broadband in the UK," February 27, 2013, https://www.gov.uk/broadband-delivery-uk.

11 Australian Government, "National Broadband Network," n.d., http://www.communications.gov.au/broadband/national_broadband_network.

12 Adam Oxford, "100MBps Plan Takes South Africa a Step Closer to Universal Broadband," ZDnet.com, October 28, 2013, http://www.zdnet.com/100mbps-plan-takes-south-africa-a-step-closer-to-universal-broadband-7000022502/.

13 Universal Service Administrative Company, "Broadband Pilot Program," n.d., http://www.usac.org/li/about/broadband-pilot/default.aspx.

14 The other G7 countries are the United States, the United Kingdom, France, Germany, Italy, and Japan.

15   CRTC, "CRTC Initiates Review of Basic Telecommunications Services for All Canadians," press release, April 9, 2015, https://www.newswire.ca/news-releases/crtc-initiates-review-of-basic-telecommunications-services-for-all-canadians-517432291.html.

16   ACORN Canada, "Internet for All: Internet Use and Accessibility for Low-Income Canadians," February 2, 2016, https://acorncanada.org/resource/internet-all.

17   Christine Dobby, "CRTC Chair Makes Strong Call for National Broadband Strategy," *Globe and Mail,* April 18, 2016.

Transmutations

Nehal El-Hadi

I see the work that I'm doing as community building
because I kind of started it – what I'm doing – as wanting
that for myself and wishing that I had that kind of
community for myself.   BEE QUAMMIE, MAY 6, 2015

I like to consider myself a storyteller.   SHIREEN AHMED, MARCH 31, 2015

I would say that I am a different perspective-giver
and a status-quo challenger. I like to ask questions,
and I like to expand our understanding of things.   IDIL BURALE, MAY 7, 2015

I'm a resource person, the connector; you have an idea,
I know a million other people that can help you with that
idea. You have an issue, I might not be able to help you, but
I know twenty other people who can support you. So that's
what I do ... I definitely see myself as a city builder. I also see
myself as a bridge builder. I see myself as somebody who
connects.   HIBAQ GELLE, MAY 15, 2015

I am a community artist. I work with the community to
cultivate arts-based projects, usually around issues of violence,
identity, and social location. As well, I do education. I'm a
public speaker. I do a lot of public-speaking engagements, and
I do different things on social media to gain people's interest
in different issues that I think are important. Mostly, I'm a
counsellor, and a hugger. I like hugging.   FARRAH KHAN, APRIL 1, 2015

I'm an arts educator ... I'm a lecturer, I'm a public researcher. I share information ... I'm a yoga teacher ... I'm a writer ... I'm an artist, so I'm a curator ... I'm still rooted in Toronto.

KIM KATRIN, APRIL 17, 2015

So artist, scholar, educator.

AMANDA PARRIS, APRIL 2, 2015

I would say my work is storytelling and documenting.

NAYANI THIYAGARAJAH, APRIL 10, 2015

I do a few different things.

SHEILA SAMPATH, MARCH 27, 2015

We been livin' through your internets.[1]

DANIEL BANGALTER, ET AL., "THE HEALER"

I like to create spaces. I think that's my favourite thing – helping to create spaces for people to connect, to create, to imagine. That's by far my favourite thing to do.

AMANDA PARRIS, APRIL 2, 2015

This task of making homeplace was not simply a matter of black women providing service; it was about the construction of a safe place where black people could affirm one another and by doing so heal many of the wounds inflicted by racial domination. We could not learn to love or respect ourselves in the culture of white supremacy, on the outside it was there on the inside, in that "homeplace," most often created and kept by black women, that we had the opportunity to grow and develop, to nurture our spirits. This task of making a homeplace, of making home a community of resistance, has been shared by black women globally, especially black women in white supremacist societies.[2]

BELL HOOKS, *YEARNING*

One of the things that really appeals to me about working online is so many of the people who are excluded from participating in most public spaces – because of class, they don't have the money, because of access, most spaces are really ableist, in all of Toronto there's very few accessible spaces, people that don't have ASL [American Sign Language], places that are not equipped for blind folks – online, you get to find all of those people. Online, you get to meet them. You get to build with them and learn about them and learn about them from them, and that is so exceptional for me in terms of the work that I do.

KIM KATRIN, APRIL 17, 2015

Far from appearing antithetical to the human organism and set of values, the technological factor must be seen as coextensive with and inter-mingled with the human. This mutual imbrication makes it necessary to speak of technology as a material and symbolic apparatus, i.e. a semiotic and social agent among others.[3]

ROSI BRAIDOTTI, "CYBERFEMINISM WITH A DIFFERENCE"

And I'm starting to have an issue with when we're saying "online versus in real life" because even with something like Twitter, I find that I make certain connections with people who I've never met and maybe will never meet in person, but those connections are just as significant if not more than some people who I see every day. So I'm feeling the same way. I'm having a bit of an issue with making that differentiation between what's real life and what's not real life. I'm starting to say "online versus off-line," so I feel you on that.

BEE QUAMMIE, MAY 6, 2015

So that's why I feel my work happens in very, very small ways. I feel like those small ways are really important and also large ways. That's why I think, for me, I like to use online tools and strategies to share that information, because that's how the small pieces can end up being disseminated largely and have meaning for a larger group without taking away from the need to have those incredibly small, personalized niche spaces.

KIM KATRIN, APRIL 17, 2015

To claim a butterfly's wings can cause a storm, after all, is to raise the question: How can we definitively say what caused any storm, if it could be something as slight as a butterfly?[4]

PETER DIZIKES, "THE MEANING OF THE BUTTERFLY"

All of this is through social media pretty much, except for some stuff that's done locally.

SHIREEN AHMED, MARCH 31, 2015

There's Torontonians that don't experience the same Toronto, essentially.

HIBAQ GELLE, MAY 15, 2015

I do think policing and police relations is one of our generation's civil rights issues. So whether you're in Toronto or Baltimore, it's a question of who gets to own the city and who gets to walk the city at any given time and not be deemed suspicious.

IDIL BURALE, MAY 7, 2015

Street harassment is really real. Racism is really, really real. So online, I feel that the negative attention I get, at least I can block it.

KIM KATRIN, APRIL 17, 2015

I was talking about how activism, before it's an external process, before it's an organizing process, is a deeply internal process, especially for racialized people, especially for women, especially for queer folks, where you have to learn to love yourself, where you have to learn to be okay with who you are to a certain extent, and then from there an extension of that to interpersonal stuff.

SHEILA SAMPATH, MARCH 27, 2015

We're safe. We're not judgmental. These are spaces that are local in Toronto we're very, very blessed to have. Because although we live in a country and a city that's hugely multicultural, are there really safe spaces for us that aren't overrun with male voices, or voices that are predominantly of white people?

SHIREEN AHMED, MARCH 31, 2015

And if people need to have a conversation, I have the ability to be, like, "Well, you're not having this conversation by yourself. These few people need to be at this table. Sometimes I have to assert myself in spaces. Sometimes I'm not really wanted in spaces. But you know what? You're not going to have this conversation without these key people." I like tension. I like awkwardness. I love bringing people to understand that it's not just you. There's a bigger conversation.

HIBAQ GELLE, MAY 15, 2015

Online self-determination can also mean one's technical, and very importantly, financial ability to represent and edit oneself and one's culture(s) online, and to decide how they will achieve online relevance/visibility/ranking without being overshadowed by more dominant national languages and/or economies.[5]

ERNESTO PRIEGO, "CAN THE SUBALTERN TWEET?"

So somebody literally found me online, through Twitter, and was, like, "Wow, your work is really, really, really important." And when I say "work," it was just a blog piece.

SHIREEN AHMED, MARCH 31, 2015

As such, internet protests may be a useful tool for black women to augment traditional approaches to advocacy and activism, allowing them to have an explicit public presence with the opportunity to influence local law enforcement practices, while also raising awareness at the local and national level ... The internet presents a unique site for studying the utility of the web for disrupting hegemonic understandings of violence against women and racial minorities for transforming the way historically marginalized groups are treated in their communities and by the criminal justice system.[6]

LAURA RAPP, ET AL., "THE INTERNET AS A TOOL FOR BLACK FEMINIST ACTIVISM"

My audience is big enough at this point that I want to make sure that I'm being most respectful to the people who have most investment in an issue and would be considered the most sensitive about it and their feelings would so often be dismissed.

KIM KATRIN, APRIL 17, 2015

But it's also where trolls live. That's what I learned.

AUSMA MALIK, APRIL 14, 2015

It got so intense, people were sending me death threats. People were telling me that I was the reason that everything was wrong in the world. Just really intense things because I was trying to explain, especially to white people, why reverse racism is not a thing and why it's really dangerous to talk about it as though it was a real experience in the world.

KIM KATRIN, APRIL 17, 2015

Whenever I talk to other people who are doing online work, I'm always, like, "Have your block game on heavy. Protect yourself while you're doing this work."

KIM KATRIN, APRIL 17, 2015

Over the past 10 to 15 years in particular, feminist spaces have been concerned with and consumed by an Ahab like quest for building and enforcing "safe space." As women of colour, who live under white supremacy, settler colonialism, hetero-normativity, capitalism and more, we know that such a place doesn't actually exist. More importantly, what we have seen over the years is that "safe spaces" usually mean excluding us. They sometimes mean using "safety" as a substitute for "never uncomfortable spaces." In this conceptualization, safety is often used as a cudgel to silence and to

ANDREA SMITH AND MARIAME KABA, "INTERLOPERS ON SOCIAL MEDIA" further marginalize.[7]

You use social media like we use all sorts of different tools as organizers. We have strategy. We want to work to this end, and we have those goals, and social media for me has always been built into that; it's been one of the tools that we use to

AUSMA MALIK, APRIL 14, 2015 movement build.

So I have an awareness of brand because I did think about that. That's part of where my brain went to, but I don't think about it too seriously because that scares me. It scares me.

AMANDA PARRIS, APRIL 2, 2015 Being a product.

Waiting for our multicultural future must include those practices of racial violence that have provided, and continue to provide, the grounds through which we anticipate a

KATHERINE MCKITTRICK, "WAIT CANADA ANTICIPATE BLACK" different future.[8]

But we cannot look at the presence of marginalized women in digital spaces without considering our oppression. What some are truly afraid of are the layers that begin to unfold if we take a more careful look at how women are using Twitter to engage with a movement they previously had trouble connecting to because of disability, interpersonal violence that limited their movement, marginalized motherhood with little support, transphobia and class. When our voices come to the fore, mainstream organizations and anti-violence movements have to come to terms with the fact that we might have a

SHAADI DEVERAUX, "WHY THESE TWEETS ARE CALLED MY BACK" different vision.[9]

But we can cheat on capitalism and take care of ourselves and take care of each other, and it's really important for us to do those kinds of things.   KIM KATRIN MILAN, APRIL 17, 2015

POSTSCRIPT (PLUS THREE YEARS)

What you just read is a textual assemblage. It presents material from other sources to construct a derived narrative. The quotes are removed (but not too far) from their original context. Composed of many voices, the new narrative presents a more complex version of the truth.

This approach to truth telling has three sources of inspiration. First, in *Culture and Imperialism,* Edward Said makes an argument for what he calls the contrapuntal reading of texts: "As we look back at the cultural archive, we begin to reread it not univocally but contrapuntally, with a simultaneous awareness both of the metropolitan history that is narrated and of those other histories against which (and together with which) the dominating discourse acts."[10]

The second inspiration was remix culture – especially mixtape production – in which previously read or heard material can be disassembled and reassembled, decontextualized and recontextualized, producing new and original meanings that are still connected to the original source material.

The third inspiration was journalist Janet Malcolm's Free Associations collages in which she transforms archival materials into works of art. Commenting on them, Hilton Als writes:

> Malcolm's desire to order the world is not so much the desire to re-create or control it as it's an exploration of its various elements – those moments of being that are no more, and that were as true and fake as anything else. Grief and fiction are the central themes of her collages; the grief is real, the images are made up out of the real stuff of grief, which is to say artifacts from the past, a desire to not let go, and are the visual representations of the will to remember even as time erodes that will, and we are no more.[11]

Using the textual assemblage as a form, I created a text made from "real stuff." "Transmutations" is based on a series of interviews I conducted with ten women of colour social justice activists in Toronto on how they are using their online activities to access their right to the city. My intention was to show how the online activities of women of colour reflect their engagement with urban space. I combined quotes from the interviews with fragments from their online public archives and more academic texts to formulate a theory of the production of presence.

The production of presence is the creation and distribution of content and stories that centre marginalized individuals and reflect the being-in-the-world of marginalized groups in deliberately authentic and representative ways. The narratives that come of the process counter dominant hegemonic discourses of everyday life by providing alternative retellings and imaginings of lived experiences. As theory in praxis, there is something alchemical about the way that actions taken in virtual space have effects in material space, a transmutation of one action into a different one. A retweet signal boosts an event into an action, such as the shutting down of Juror B37, a case where online lobbying resulted in the withdrawal of a book deal from a juror in the George Zimmerman case. Similarly, the hashtag #YouOKSis, originated by Feminista Jones, contains the power to transform urban space and make city streets safer for women who may be harassed walking down them. It relies on the power of witness, exercised by a simple "You OK, Sis?" More locally, Black Lives Matter Toronto used online organizing and connecting to an international network of Black Lives Matter movements to arranged for the occupation of urban space. This protest closed down the Allen Expressway for two hours and camped outside Toronto Police Headquarters for fifteen days.

The women whose voices appear in "Transmutations" have continued to build their networks and move from strength to strength. And, as always, I continue to be grateful to them for what they chose to share with me, and I remain inspired by them.

I am not a techno-optimist. My research and work take place in the wake of the online streaming, reproduction, and dissemination of racialized trauma and violence. And yet, in revisiting these interviews, I was reminded of what filmmaker John Akomfrah refers to as the digitopic (a portmanteau of "digital" and "utopic") possibilities of technology – the "moments when demands for the impossible became the harbingers of new modes, new relations, new systems for manufacturing and accelerating the indexical implications of the moving image."[12]

Online and off-line worlds and lives overlap – there is no false dualism that excludes one in favour of the other. The internet facilitates the creation and management of an online presence, and by performing politicized identities, women of colour access communities of support, engage in activities that enable them to accumulate social capital, and effectively channel and strengthen their communities' voice(s). Performing political identities online circumvents off-line situations and conditions that may effectively silence the voices of those marginalized from the mainstream; it facilitates not only inclusion in an abstract sense but also real material impacts on city-building processes.

I believe in the new and radical possibilities of the internet. It disrupts space and time in ways that we don't fully understand yet but that are being revealed. As Vincanne Adams, Michelle Murphy, and Adele E. Clarke have written: "Anticipation has long been a component of political practice."[13] I conducted the interviews and composed "Transmutations" in anticipation of that understanding.

NOTES

1 Daniel Bangalter, Otis Jackson Jr., Malcolm McLaren, and Erica Wright, "The Healer," on Erykah Badu's *New Amerykah Part One (4th World War)*, Universal Motown, 2008. Erykah Badu, Daniel Vangarde, Otis Jackson Jr., Malcolm McLaren

2 bell hooks, *Yearning: Race, Gender, and Cultural Politics* (New York: Routledge, 1992), 42.

3 Rosi Braidotti, "Cyberfeminism with a Difference," 1996, https://disability studies.nl/sites/disabilitystudies.nl/files/beeld/onderwijs/cyberfeminism_with_a_difference.pdf.

4 Peter Dizikes, "The Meaning of the Butterfly," *Boston Globe*, June 8, 2008.

5 Ernesto Priego, "Can the Subaltern Tweet?," *University of Venus* (blog), June 28, 2010, https://uvenus.org/2010/06/28/can-the-subaltern-tweet/.

6 Laura Rapp, Deeanna M. Button, Benjamin Fleury-Steiner, and Ruth Fleury-Steiner, "The Internet as a Tool for Black Feminist Activism: Lessons from an Online Antirape Protest," *Feminist Criminology* 5, 3 (2010): 244–62.

7 Andrea Smith and Mariame Kaba, "Interlopers on Social Media: Feminism, Women of Color and Oppression," Truthout, February 1, 2014, http://www.truth-out.org/news/item/21593-interlopers-on-social-media-feminism-women-of-color-and-oppression.

8 Katherine McKittrick, "Wait Canada Anticipate Black," *CLR James Journal* 20, 1–2 (2014): 243–49.

9   Shaadi Deveraux, "Why These Tweets Are Called My Back," *New Inquiry,*
    December 19, 2014, http://thenewinquiry.com/essays/why-these-tweets-are
    -called-my-back/.

10 Edward Said, *Culture and Imperialism* (New York: Vintage Books, 1993), 60.

11 Hilton Als, Free Associations, *New York Review of Books,* January 8, 2012,
    https://www.nybooks.com/daily/2012/01/08/free-associations-collage
    -janet-malcolm-hilton-als/.

12 John Akomfrah, "Digitopia and the Spectres of Diaspora," *Journal of Media
    Practice* 11, 1 (2010): 24–25.

13 Vincanne Adams, Michelle Murphy, and Adele E. Clarke, "Anticipation:
    Technoscience, Life, Affect, Temporality," *Subjectivity* 28, 1 (2009): 248.

Security and Surveillance

Digital Borders and Urban Worlds

Stephen Graham

WE LIVE IN A WORLD where the use of digital surveillance to construct borders within and between contemporary cities is now the norm. Our planet faces, as Asnezar Alsayyad and Ananya Roy argue, a kind of (computerized) medieval modernity that has produced forms of citizenship that are anchored in urban enclaves and fundamentally about protection. These new forms challenge the idea of modern citizenship being constituted through individual rights embedded in and protected by the nation-state. Rather than using the term "medieval" to mean a simple reversal of Enlightenment notions of progress and a return to societal backwardness, Alsayyad and Roy suggest something altogether more subtle – and convincing.

Within the transnational urban geographies of capitalism, Alsayyad and Roy see the coexistence of modern nationalism, medieval enclaves, and imperial brutality.[1] Nation-states don't simply wither away under the force of some globalized future. Rather, new camp-like enclosures and privatized circulations erupt as archipelagos within, through, and between what are conventionally understood as cities and nations. These enclaves demarcate spaces and zones deemed liberal. They are spaces and zones that require policing and militarized force from those deemed illiberal. Their existence radically complicates "the whole issue of progress and backwardness, the modern and pre-modern."[2] They also force us to interrogate the deployment of common teleologies that suggest that the barbarian or Orientalized Other inhabits the "savage past" in urban sites within the present world.[3]

The events of 9/11 accelerated processes of urban securitization and fragmentation that began long before 2001. Cities and ideas of citizenship have been progressively reorganized and revised based on provisional (rather than absolute) notions of mobility, rights, and access and on preemptive notions of suspicion, guilt, riskiness, vulnerability, and value. As in medieval times, the result seems to be that the modern city has become a honeycomb of jurisdictions, a medieval body of "overlapping, heterogeneous, non-uniform, and increasingly private memberships."[4] The modern city is permeated by new tracking, targeting, and access-control technologies and by the discourses and tactics of securitization.

Given these developments, it is tempting to adopt an apocalyptic tone. But keep in mind that borders are permeable and bordering strategies are inevitably contradictory, particularly in large cities. The emerging systems of these simultaneously digital and medieval enclaves embody technophiliac dreams of perfect ordering, perfect surveillance, and perfect power, but they inevitably fail to exert the level of geographical and social control necessary to fulfill the fantasies that drive them. In large, fast-growing cities, fortressed enclaves are often surrounded, and overwhelmed by, the mass and pulse of urban life. The sheer density and unpredictability of city living often batters down simple imaginings or strategies of boundary enforcement. Moreover, notions of security in this age of ubiquitous borders are often tenuous at best, even for those organizing or benefitting from the drive towards securitization. Maintaining borders is messy, time-consuming, and expensive work, and the complex assemblages through which securitization operates are highly precarious. They often simply address the symptoms, rather than the causes, of the spiralling insecurities faced by the world's burgeoning urban poor, in societies driven to ever-greater extremes of hyperinequality by the faltering systems of neoliberalization.

Imaginings and fantasies of perfect control and the absolute separation of the risky and the risk-free, of so-called security events and normality, remain just that: imaginings and fantasies. Like ideas of robotic warfare, these discourses are shot through with technological fetishism and dreams of omniscience and all-powerful control. They serve as a cover for myriad improvised and messy efforts to assert control, efforts that are strung out across great distances. Even with rising efforts to integrate separate surveillance systems through data mining, fusion centres, algorithmic closed-circuit television, biometrics, and the like, it is impossible to create an all-seeing Big Brother or a single global panopticon. Rather, we have many Little Brothers – an omnopticon that encompasses multiple surveillance systems of diverse scope, scale, effectiveness, and reach that sometimes interact but very often – despite the hype – do not.

Technosocial borders are always prone to technological breakdown, errors, and unintended effects. Rather than all-seeing omniscience, control rooms are marked by complex practices of improvisation that are extremely fine-grained, contingent, and socially messy. Technological dreams fail when the technology breaks down or fails or when operators fail to deal with the complexity of the system. Beyond discourses of perfect control and absolute omniscience, "the geometry of control is never complete."[5]

In addition, performances of security are about more than the policing of purported risks. Francisco Klauser has pointed out, for example, that the massive systems of temporary fortification that surround events such as the Olympics or the World Cup are also attempts to showcase and promote on a global scale the latest high-tech security solutions or impose particular brand cultures in "clean" cities.[6]

Finally, all borders and borderings are always in tension with everyday attempts at transgression and resistance. Are cities, then, in the process of being reconstructed as little more than a series of interconnected camps organized through militarized and surveilled passage points, places where all presences, and circulations, are prescreened and preapproved by continuous electronic calculation? In this context, what becomes of the right to the city and the politics of urban citizenship? Are they possible in a world of ubiquitous borderings that render urban life increasingly passive, consumerized, surveilled, and algorithmically marshalled? Will these trends fatally undermine the roles of cities as the main centres of political, cultural, social, and even economic innovation? Is ubiquitous bordering as much the result of industrial policy as it is a response to real threats? This second range of questions raises concerns about the relationship between bordering and constructions of the Other.

Above all these concerns, caveats, and crises, we must ask: What would a counterpolitics of security look like, one that actually addresses the real risks and threats that humankind faces in a rapidly urbanizing world, a world prone to resource exhaustion; spiralling food, energy, and water insecurity; biodiversity collapse; hyperautomobilization; financial crises; and global warming? What would it look like if it addressed these risks and threats from a cosmopolitan rather than a xenophobic and militaristic point of view? What would it look like if the human, urban, or ecological aspects of security were foregrounded, rather than the tawdry machinations and imagineering that surround constellations of states and transnational corporations, constellations that are integrated through the dubious and corrupt practices of a burgeoning security-industrial-military complex?

NOTES

1   Aznezar Alsayyad and Ananya Roy, "Medieval Modernity: On Citizenship and Urbanism in a Global Era," *Space and Polity* 10, 1 (2006): 17.
2   Ibid., 17.
3   Ibid.
4   James Holston and Arjun Appadurai, "Introduction: Cities and Citizenship," in *Cities and Citizenship*, ed. James Holston (Durham, NC: Duke University Press, 1999), 13.
5   Michael Shapiro, "Every Move You Make: Bodies, Surveillance, and Media," *Social Text* 83 (2005): 29.
6   Franciso Klauser, "FIFA Land TM: Alliances between Security Politics and Business Interests for Germany's City Network," in *Architectures of Fear,* ed. Centre for Contemporary Culture Barcelona (Barcelona: CCCB, 2007), retrieved from http://publicspace.org/ca/text-biblioteca/eng/b035-fifa-land -2006-tm-alliances-between-security-politics-and-business-interests-for -germany-s-city-network.

Audre Lorde's File and June Jordan's Skyrise

Simone Browne

> What kind of schools and what kind of streets and what kind
> of parks and what kind of privacy and what kind of beauty and
> what kind of music and what kind of options would make love
> a reasonable, easy response?
>
> June Jordan,
> *Civil Wars: Observations from
> the Front Lines of America*

THE DECLASSIFIED FILE that the FBI compiled for more than eighteen years on Audre Lorde is short, as far as these things go. It is only twenty pages in length. Some handwritten memoranda. Some duplicates of other pages. A note on her poem "Prologue," published in 1972 in the journal *Freedomways*. A few other matters. In the file she is indexed as "AUDREY LORDE, aka Audrey Lourdes, SECURITY MATTER – C." "C" stands for communist.[1]

Lorde first appeared in a memo dated September 9, 1954, from the legal attaché, or legat, in Mexico City, the FBI's oldest foreign office. An informant, whose name is redacted, told the bureau that they had met Audre Lorde in Oaxaca, Mexico, in June of that year. She is described as a "twenty-year-old negro girl" who says "she resides in New York on or near 7th street." The informant is said to have told the FBI that Lorde had been in Mexico since March 1954, studying at the Polytechnic Institute in Mexico City. The memorandum notes that when Lorde is in Cuernavaca, Mexico, she lives at a residence where "a number of known American communists have from time to time stayed at."

This September 9, 1954, memorandum also notes that the inform-ant stated that although Lorde "claimed to not be a member of the Com-munist Party, she did at least give the impression of following the Party line in conversation." In other parts of the FBI file, Lorde is described either as a student at Hunter College in New York City or as "a teacher of Black Literature at John Jay College." Another memo in the file, dated June 17, 1957, and sent to the FBI's New York field office from Special Agent Joseph T. Quigley, states that Lorde was interviewed at her home for approximately twenty-five minutes. She is described as "cordial" and "friendly throughout the interview," and when she was asked to give further consideration to cooperating with the bureau, she is said to have responded that "she was sure that she wouldn't change her mind."

These declassified files must be read with their original intent in mind. They were teletyped messages sent back and forth between field offices and headquarters that provided interpretations of the bureau's surveillance of, and preoccupation with, the captioned subject. It is, of course, a question of who the intended reader is. Often, when a file is declassified, much of the original text is redacted with black bars or white blocks, concealing the censored text. But redaction, as Anjali Nath suggests, "does not negate the possibilities of interpreting and making meaning from these files."[2] Although the redactions made to the de-classified file on Audre Lorde are few, I highlight this June 17, 1957, memo for the ways her refusal to cooperate comes through. As the memo notes, "She would not report on her friends." It states further that Lorde would not report on "anyone else she knew." Cordial enough, Lorde would not cooperate with the FBI and "wondered why the FBI was in-terested in her for that work."

In *Zami: A New Spelling of My Name*, Lorde tells us just as much when she writes of another time that the FBI visited her at her New York City apartment: "The day before the FBI agents stood in my doorway, Eva Perón had died in Argentina. But somehow *we* were a threat to the civilized world." In the height of the McCarthy era, Lorde knew enough to "not to let them past my door. They stood outside, stupid and male and proper and blond and only a little bit threatening."[3] She would not report on her friends.

June Jordan likewise meditated on the FBI, the CIA, resistance, and, as she put it, "the history of the terrorized incarceration of myself." She does so in "Poem about My Rights," but it is in *Civil Wars: Observations from the Front Lines of America*, that she writes about her desire for safety, a desire that led to her creative reimagining of Harlem, New York, by way of a "collaborative architectural redesign" with Richard Buckminster "Bucky" Fuller.[4]

July 18, 1964, marked the beginning of six days of rioting in Harlem following the killing of James Powell, a fifteen-year-old black boy, by a police lieutenant. That same year, Jordan proposed to fellow architect Fuller that they "work together to design a three-dimensional, an enviable, exemplary life situation for Harlem residents who, otherwise, had to outmanoeuvre New York City's Tactical Police Force, rats, a destructive and compulsory system of education, and so forth, or die."[5]

Jordan named their plan "Skyrise for Harlem," and it addressed what Jordan called "a main worry," namely, that "too often, urban renewal meant Negro removal." Instead, Skyrise, for Harlem, would be an "environmental redesign as a form of federal reparations to the ravaged peoples of Harlem" in the form of "beautiful and low-cost shelter integral to a comprehensively conceived new community for human beings."[6] The plan for a New Harlem included expansive walkways, parks, and other communal green spaces; a new bridge to New Jersey; and fifteen large apartment towers with living units for up to half a million people. Rather than streets laid out in a grid-like fashion, they would seek "life-expanding design for a human community" in the form of "as many curvilinear features of street patterning as possible."[7] When the proposal was published in the April 1965 edition of *Esquire* magazine, it was renamed "Instant Slum Clearance." Although Jordan is credited as the article's author (but with the byline "June Meyer," her married name), only Fuller is credited for the architectural design of New Harlem.

• • • • •

The FBI's surveillance of Audre Lorde came right at the commencement of the bureau's Counterintelligence Program (COINTELPRO), which was designed to discredit and disrupt various domestic political organizations through disinformation, infiltration, violence, and other means. Although COINTELPRO is, officially, no more, this type of targeted surveillance has been ongoing. Think of the New York Police Department's (NYPD) now disbanded Demographic Unit, which engaged in profiling, spying on, and tracking Muslims at mosques and other so-called hot spots, the FBI's creation of the Black Identity Extremist as a new category of threat, and the NYPD's and FBI's surveillance of the Black Lives Matter movement, to name just a few.[8]

June Jordan's desire for safety, security, and something other than having to outmaneuver New York City's Tactical Police Force, as she put it, is still relevant. It is a desire for a type of city life that counters a capitalism that is unrestrained yet hinged to repressive policing tactics in the form of, for example, the NYPD's Omnipresence strategy. Police project floodlights or flashing lights from their cruisers onto so-called hot spots or so-called high-crime neighbourhoods, exposing their occupants to constant, high-intensity light, loud noises and putrid smells from portable generators, and other harmful effects such as severe glare and the suppression of the production of melatonin.[9]

Omnipresence is part of a collection of policing tactics (including fusion centres, predictive policing, acoustic gunshot detection, body cameras, and cell-site simulators) that, as Joshua Scannell writes, leverage "algorithms to bathe a city in electric light."[10] Surveillance and policing tactics such as these, along with their antecedents, often force people to seek do-it-yourself solutions for community self-defence, solutions such as Audre Lorde's refusal to cooperate and June Jordan's demand for the kind of options for city life that would make "love a reasonable, easy response."

NOTES

1 Federal Bureau of Investigation, 2018. Audre Lorde, case file, 100-NY-122142, https://archive.org/details/AudreLordeFBI/page/n15/mode/2up.

2 Anjali Nath, "Beyond the Public Eye: On FOIA Documents and the Visual Politics of Redaction," *Cultural Studies/Critical Methodologies* 14, 1 (2014): 22.

3 Audre Lorde, *Zami: A New Spelling of My Name* (New York: Crossing Press, 1982), 121.

4 June Jordan, *Civil Wars: Observations from the Front Lines of America* (New York: Touchstone, 1995), xiii.

5 Jordan, *Civil Wars*, xi. See also Alexis Pauline Gumbs, "June Jordan and the Black Feminist Poetics of Architecture," *Plurale Tantum*, March 21, 2012.

6 Jordan, *Civil Wars*, 24.

7 Ibid., 27.

8 Jana Winter and Sharon Weinberger, "The FBI's New U.S. Terrorist Threat: 'Black Identity Extremists,'" *Foreign Policy*, October 6, 2017, https://foreign policy.com/2017/10/06/the-fbi-has-identified-a-new-domestic-terrorist

-threat-and-its-black-identity-extremists/; and New York Civil Liberties Union, "Court Rejects Secrecy Loophole, Rules NYPD Must Respond to Records Request Regarding Surveillance of Protestors," NYCLU, January 14, 2019, https://www.nyclu.org/en/press-releases/court-rejects-secrecy -loophole-rules-nypd-must-respond-records-request-regarding.

9   R. Joshua Scannell, "Electric Light: Automating the Carceral State during the Quantification of Everything" (PhD diss., City University of New York, 2018), and Joseph Goldstein, "Stop-and-Frisk Ebbs, but Still Hangs Over Brooklyn Lives," *New York Times*, September 19, 2014.

10   Scannell, "Electric Light," 5. Fusion centres are designed to promote information sharing at the federal level between agencies and law enforcement. Cell-site simulators are devices disguised as cellphone towers that make phones within a certain radius connect to the device rather than to a regular tower.

Policing the Future(s)

R. Josh Scannell

OVER THE PAST decade, technical innovations in the capacity to collect, store, and analyze massive amounts of data have created a global market – valued in the hundreds of billions of dollars – for what are commonly called smart cities. Smart cities are municipalities that have partnered with private technology firms to analyze enormous amounts of digital information produced and collected rapidly by the city and its inhabitants.[1] Depending on the program and its reach, the amount of information analyzed at any given moment might be on the order of several petabytes (one petabyte equals one million gigabytes).[2] Some cities, such as Rio de Janeiro, have built fusion centres, control rooms for managing digital information designed to coordinate information across dozens of municipal departments – from traffic flow to emergency weather alerts.[3] Others, such as New Songdo in South Korea, have been built from the ground up to be completely digitally interactive, essentially rendering the city as a massive network of digital sensor arrays.[4]

In theory, access to such simultaneously vast and specific data allows urban municipalities to govern with maximum efficiency by predicting, and correcting for, social disturbances before they happen.[5] In practice, the adoption of smart-city technology raises worrying questions about the capacity of states and infrastate organizations to control their citizens. This is particularly true of the branch of urban management that has most consistently embraced and implemented these technologies: criminal justice.

New York City provides an excellent case study of the questions raised by the deployment of smart-city technologies and predictive policing as putatively neutral best practices for governance. New York is, and has historically been, one of the most racially segregated, economically unequal cities in the United States. Racist urban-renewal projects, post-industrial economic collapse, white flight, market deregulation, gentrification, and a host of other factors have combined to heavily articulate poverty and race within the urban fabric. These processes have historically played out in tandem with the criminalization of poverty, racialization, and queer-phobic policing in the city. New York's deployment of law enforcement (particularly after the introduction of broken-windows policing) as a means of maintaining what Cedric Robinson has called racial capitalism, has produced, for black and brown populations especially, an ongoing crisis of the criminalization of daily life.[6] It is in this context that the city has built, at massive expense, a data analytics system designed to collate the entirety of digital surveillance data generated by the City of New York to "neutrally" and "rationally" predict crime and criminality.

On August 8, 2012, the mayor's office announced the unveiling of "new, state-of-the-art law enforcement technology that aggregates and analyzes existing public safety data in real time to provide a comprehensive view of potential threats of criminal activity."[7] Called the Domain Awareness System (DAS) and developed as a private-public partnership between Microsoft Corporation and the New York Police Department, the system "pools existing streams of data from live camera feeds, 911 calls, mapped crime patterns and more to help officers prevent crimes and respond faster."[8] In theory, this system gives Microsoft technicians the processing power to constantly collect all of the digital information generated by the city and render it intelligible for NYPD command, in something like real time.

The system's unprecedented networking of closed-circuit surveillance cameras is particularly unnerving. At the time of the system's unveiling, New York City already had approximately three thousand CCTV cameras, most concentrated in the financial nerve centres of Lower Manhattan and Midtown. By 2015, that number had increased to over seven thousand, dispersed throughout the boroughs, though still concentrated in Manhattan. Most of the camera feeds available to DAS analysts are owned by the public, but various stakeholder organizations such as the Federal Reserve, the Bank of New York, Goldman Sachs, Pfizer, and Citigroup agreed to allow the NYPD to access their own surveillance systems and incorporate these feeds into the DAS.[9] The system allows the department to access both current feeds and stored and archived streams; for instance, if officers receive a tip about a suspicious package left at a particular location, technical staff can quickly tap into archived feeds and identify the person who left it. By linking video feeds to other information streams, the city imagines the capacity to "track where a car associated with a suspect is located, and where it has been in past days, weeks or months."[10]

Closed-circuit cameras make up only a fraction of the surveillance apparatus networked through the DAS. Much of the data collected and analyzed by the NYPD are, following the Edward Snowden National Security Agency (NSA) revelations, seemingly mundane: cellular metadata, arrest records, smart cameras, licence-plate trackers, social-media feeds, and 911 calls. These are but a few of the data sources now collected, sorted, and analyzed by Microsoft technicians.

Some information streams are more specific. The Department of Homeland Security largely funded the system's $40-million development under a counterterrorism mandate. The system's Public Security Privacy Guidelines refer to environmental data as "data collected by devices designed to detect hazards related to potential terrorist threats, or to respond to terrorist attacks." It is unclear what this means, exactly, except in one instance: radiation sensors. Installed presumably to detect the hypothetical dirty bomb in a suspicious package, the sensors are perceptive enough, and the analytics system sophisticated enough, to "quickly identify whether the radioactive material is naturally occurring, a weapon, or a harmless isotope used in medical treatments."[11] The system is capable of knowing whether a passing body on the street has recently had chemotherapy.[12]

These surveillance capabilities are among the more notable aspects of the system's crisis-response capabilities. But, as Jasbir Puar has pointed out, "technologies of surveillance [are] not only responsive and thus repressive, but also pre-emptive and thus productive."[13] And, true to form, a major selling point of DAS is its pre-emptive capacity in relation to so-called crime. The system draws on rich data sources to generate heat maps of likely criminal activity. Based on meteorological images that use visual cues, like colour, to map probable temperatures onto space, these heat maps are organized taxonomically, spatially, and chronologically. That is, the maps predict (1) the type of criminal activity likely to occur, (2) the space in which the criminal event is likely to unfold, and (3) the timing of the event. The maps, in other words, morph. The idea is that, armed with knowledge of where and when a supposed crime will occur, precincts will be able to pre-emptively distribute police officers in space and time to prevent the criminal event from happening.

Should an incident occur, the officers will be equipped with fore-knowledge of what type of crime will happen (misdemeanour, felony, property offence, violent crime, drug-related incident, etc.) and armed with the proper array of weapons to contest and contain it.[14] This approach effectively legitimizes departmental decisions to deploy overwhelming force to particular areas and situations, without effective oversight from civilians. If the DAS indicates to the police department that heavy fire-power is needed in any given situation, it is incumbent on the department to arm officers and act accordingly. It locks the logic of the police as an occupying body into the practices and protocols of how to engage with targeted urban spaces. That these spaces are themselves the products of racist municipal policies – from zoning to distribution of public resources – compounds the punitive logic by which racialized communities are rendered, violently, apart from, and exceptional to, the rest of the city.

While the DAS's surveillance capacities are in line with practices that critics usually target when they worry about eroding civil liberties (for example, the NSA's data-collection program), it's the pre-emptive capabilities that are particularly exciting to NYPD brass and private stakeholders.[15] When pushed by City Council in May 2014 to explain the NYPD's $4.71-billion plus budget for DAS, police chief Bill Bratton vigorously defended predictive policing. He argued that that predictive policing is here to stay and that Microsoft and the NYPD "are beginning to write algorithms that identify in a real-time way paths of criminal activity."[16] Moreover, he stated that the next stage of tech development would involve using the city's proprietary New York City Wireless Network (a "mission-critical broadband infrastructure") to make the system mobile. Eventually, the NYPD hopes to have DAS-networked Microsoft tablets in the hand of every police officer, presumably to eliminate the data choke points and temporal drag of the human chain of command.[17] Given the public-private partnership between the city and Microsoft, this raises the troubling prospect that the data streams used by the police department to determine their policies will be generated by a for-profit institution whose algorithms are legally firewalled from the public they claim to study. Microsoft's bots articulate the horizons of possible police actions.

The NYPD, for its part, is interested in its units being able to respond in real time to so-called crimes that Microsoft's algorithms think will happen in the future. For practical purposes, this means legitimizing pre-emptive police action against likely sites of criminal activity and the bodies most likely to commit it. More expansively, the city is transformed into an endless topography of possible criminality.

Perversely, the more unbounded and expansive the criminality, the more valuable the city becomes. In addition to being a tool for policing, the DAS is a capital generator for the city. Microsoft owns the technology, but the city gains 30 percent of the revenue of every copy sold. Microsoft has been aggressively marketing its product to cities around the world as "Aware." The arrangement was, from the beginning, envisaged and sold as an alternative financing model. At DAS's unveiling, Mayor Bloomberg said, "Citizens do not like higher taxes, so we will (find other revenue outlets) ... I hope Microsoft sells a lot of copies of this system, because 30 percent of the profits will go to us."[18] Under the city's agreement with Microsoft, the practice of data-driven policing is itself profit generating. The city, in other words, now has a financial incentive to pursue broken-windows policing as a logic of the state. Its capacity to operate is increasingly dependent on deploying policing violence on its denizens.

Critical discussions of DAS and systems like it almost always invoke either George Orwell's *1984* or Phillip K. Dick's "Minority Report." These works can be useful touchstones, but the reality is that new modes of pre-emptive policing have their roots in the last few decades of existing police strategies (such as the broken-windows theory and CompStat, which was driven by NYPD police chief Bill Bratton in the 1990s) and the increasing intermingling of the state and private capital under neoliberal paradigms of governance.

Broken windows was developed as a controversial theory of crime prevention in 1982 and embraced enthusiastically by New York's mayors and police commissioners since the late 1980s. Its basic premise is that minor acts of disorder encourage a communal divestment from space, which leads to generalized social disorganization and more serious acts of harm; in other words, it offers a trickle-up theory of crime. To pre-empt this process, police officers are ordered to rigidly enforce low-level "quality of life" city ordinances (selling loose cigarettes, tagging a wall, loitering, drinking on a stoop, etc.) by profligately charging people for minor violations. If a neighbourhood is visibly ordered, the logic goes, then residents are more likely to take pride in it and try to maintain it. This theory offers a speculative logic of policing. It demands that officers act aggressively in the present to ward against possible, emergent threats in the future.[19]

CompStat is technology designed to rationalize policing by quantifying police work and using geographic information systems to map criminal activity. Commanders meet regularly to compare criminal statistics for their precincts and maps of past criminal hot spots. They then formulate strategies for best utilizing their forces to drive down future criminal activity. The system is designed to streamline the project of crime fighting into a dispassionate accounting of statistics and numbers. It depends largely on broken-windows–style community policing to function in that there must be a steady stream of crime data flowing from the street to the boardroom. Granular crime data have historically meant large-scale and largely racialized arrests, charges, fines, and other criminalization practices.[20]

DAS, then, is a sort of hybrid of these two systems; it has the massive computing power of big data with urban informatics thrown into the bargain. The logic that undergirds the system is threefold. First, crime is mathematizable – it is a sort of natural phenomenon, like weather, that can be neutrally predicted and anticipated. This approach occludes not only the social construction of crime statistics but also the social construction of crime and the urban. Second, urban space is a laboratory for government. This is less a narrow question of privacy than a broad one of what, and whom, the city is for. Third, policing is an end in itself. Broken windows has been the only overarching theory of policing for thirty years. The DAS, by mathematizing crime statistics and trends developed under the broken-windows regime, effectively naturalizes a controversial theory on what governance is. Broken-windows theory has driven a massive and devastating criminalization of the city's nonwhite and poor populations. It has codified white paranoia of nonwhite bodies as legitimate and as the basis for policing crime.

Although Microsoft and the NYPD seek to portray this system in beneficent terms, the picture of the urban that these practices conjure is unremittingly bleak and totalitarian. The system enshrines the pre-emptive criminalization of already-marginalized urban denizens as neutral, necessary, and smart police praxis. It also renders the system and its underlying logic as profitable for transnational capital: less criminalization (but not less "crime") is bad for business. It is particularly disturbing, then, that this system has been successfully marketed to cities not only in the United States (Long Beach, Newark, and Boston, among others) but also around the world (São Paolo recently launched the technology, rebranded as Detecta). This transnational success suggests a truly frightening neoliberal consensus that blends carceral logic, prison industrial complexes, and disruptive technology in the name of making cities safer and tech companies more profitable. That this profitable safety must be extracted from the criminalized bodies of the least powerful tends to be left out of the advertising copy.

NOTES

1   Adam Greenfield, *Against the Smart City* (Seattle: Amazon Digital Services, 2013).
2   Rob Kitchin, *The Data Revolution: Big Data, Open Data, Data Infrastructures and Their Consequences* (New York: Sage, 2014).
3   Natasha Singer, "Mission Control, Built for Cities," *New York Times*, March 3, 2012, https://www.nytimes.com/2012/03/04/business/ibm-takes-smarter-cities-concept-to-rio-de-janeiro.html.
4   Orit Halpern, Jesse LeCavalier, Nerea Calvilla, and Wolfgang Pietsch, "Test-Bed Urbanism," *Public Culture* 25, 2 (2013): 272–306.
5   Stephen Goldsmith and Susan Crawford, *The Responsive City: Engaging Communities through Data-Smart Governance* (New York: Jossey-Bass, 2014).
6   Cedric Robinson, *Black Marxism: The Making of the Black Radical Tradition* (Chapel Hill: University of North Carolina Press, 1983).
7   Stu Loeser and Marc Vorgna, "Mayor Bloomberg, Police Commissioner Kelly and Microsoft Unveil New, State-of-the-Art Law Enforcement Technology ...," *News from the Blue Room*, March 3, 2012, https://www1.nyc.gov/office-of-the-mayor/news/291-12/mayor-bloomberg-police-commissioner-kelly-microsoft-new-state-of-the-art-law.
8   Ibid.
9   Neal Ungerleider, "NYPD, Microsoft Launch All-Seeing 'Domain Awareness System' with Real-Time CCTV, License Plate Monitoring," *Fast Company*, August 8, 2012, http://www.fastcompany.com/3000272/nypd-microsoft-launch-all-seeing-domain-awareness-system-real-time-cctv-license-plate-monito.

10 Loeser and Vorgna, "Mayor Bloomberg."

11 Ibid.

12 Ungerleider, "NYPD, Microsoft Launch."

13 Jasbir Puar and Lewis West, "Jasbir Puar: Regimes of Surveillance," *Cosmologics*, December 12, 2014.

14 Laura Nahmias and Miranda Neubauer, "NYPD Testing Crime-Forecast Software," *Capital New York*, July 8, 2015.

15 Colleen Long, "NYPD, Microsoft Create Crime-Fighting 'Domain Awareness' Tech System," *Associated Press*, April 22, 2013.

16 Carol Schaefer, "Bratton Calls Marijuana Decriminalisation 'Major Mistake,' May Not Realize it Already Exists," *Gothamist*, May 20, 2014, https:// gothamist.com/news/bratton-calls-marijuana-decriminalization-major -mistake-may-not-realize-it-already-exists/.

17 Pervais Shallwani, "New NYPD Tablets Help Fight Crime," *Wall Street Journal*, June 26, 2014.

18 Ungerleider, "NYPD, Microsoft Launch."

19 Bernard Harcourt, *Against Prediction: Profiling, Policing and Punishing in an Actuarial Age* (Chicago: University of Chicago Press, 2007).

20 Ibid.

● Policing Borders through Sound

Anja Kanngieser

IT CAME FROM ALL around. The weight of helicopters chopping the sky with their blades overhead, the repetitive whips undercut by cracking drones oscillating as they moved from one location to the next, hovering, then peeling away. The loud bursts of flashbang grenades. The atmosphere of war made in the air to etch out zones of militarization, as vertical as they were horizontal. The sound of missiles and machines, of infrastructural collapse, carried through the air in waves.

Waveforms of sound make it almost impossible to escape, as they are felt as much as heard. Writing on the affective force of sound, Steven Goodman argues that sound warfare affects how "populations feel – not just their individualized, subjective, personal emotions, but more their collective moods or affects." For Goodman, "sound contributes to an immersive atmosphere or ambiance of fear and dread."[1] Whether on land or sea, sound has the capacity to move with significant force. When used in warfare (often based on the intensity of volume or frequency), it is particularly difficult to avoid because sound waves are vibrational, being air- and matter-borne. The vibrational nature of sound also means that it acts upon entities regardless of whether those entities are conscious of it (for instance, in the case of ultra- and infrasounds). One cannot truly shut out sound.[2] When it is not heard, it still penetrates skin and bodily cavities, organs, and cells.

THE SONIC GOVERNANCE OF SPACE

The use of sound in the policing of public spaces and border zones is becoming more prevalent, especially with technologies that rely on loud volumes to disperse or contain populations. Most common among these is the long-range acoustic device (LRAD) produced by the American Technology Corporation – a "high-intensity directional acoustic hailer designed for long-range communication and issuing powerful warning tones" with a reach of up to three thousand metres, depending on the model.[3] It works by transmitting a high-frequency (above one kiloherz) undulating siren up to 151 decibels within a one-metre range (equal to the sound of a gunshot), and its reported effects are temporary loss of hearing, nausea, and dizziness, even for those wearing hearing protection.[4]

The LRAD is a descendent of US military loudspeakers popular since the end of the 1960s. While it was developed in a weapons program, differing views exist on the labelling of the device. American corporations categorize it as commercial; the Norwegian military categorizes it as a weapon. "One effect of this classification [as a loud hailer]," writes Jürgen Altmann, is that it requires "no legal review of compatibility with the international law of warfare."[5] Obviously, disassociating the LRAD from the rule of international law has important consequences for what the device may do, how it is used, and whether officials pay attention to its effects on human bodies.

Initially deployed by the US Navy in 2003, the LRAD was seen as a means to intervene in foreign embarkation; its far reach meant that it could be used to effectively establish a wide exclusionary zone. Over the past decade, it has been mainstreamed into logistical security, particularly in the maritime environment to aid antipiracy efforts along East Africa's coast and in the Indian Ocean.[6] The British Merchant Navy, the US navy, and commercial ocean liners and cargo ships have all adopted LRAD devices for defence against unsolicited boarding, to greater and lesser success. This lesser success has been one reason that producers have promoted the LRAD as an early-warning system rather than a weapon – one that is "highly effective in determining intent and creating standoff and safety zones in piracy applications" but unable to stop raids entirely.[7] The LRAD has been used with more effect on land than on sea. Commandeered by the US military in occupied Iraq to target both military and noncombatant populations, it was quickly adopted into policing practices, particularly during civil demonstrations.[8] From 2007, it began to appear more frequently as a tactic of less-lethal law enforcement in the United States, China, the Middle East, and Europe.

This use of military-grade technologies for domestic crowd control signals not only the extent to which lines between national and state governance are now blurred, but also the extent to which the erroneous claim that less-lethal weapons are a soft option for police response has taken hold. The LRAD is an area-of-effect weapon that targets geography rather than individuals, simply because of how sound penetrates space. The level of pain induced by the LRAD can be highly incapacitating, and it can cause permanent hearing loss. Furthermore, it envelops everyone and everything within range of its crossfire, regardless of what they are doing. Given the relative newness of the device's deployment in civilian contexts, some have argued for independent research and policies on "safe use."[9] Unlike the negligible function of its maritime counterparts, the potential of the LRAD as part of a suite of less-lethals (including tear gas, pepper spray, flashbangs, and lasers in the prevention of disorder in contemporary "threat" environments) has been noted. This is especially so within the narratives of risk associated with climate catastrophe such as increased migration, resource shortages, and austerity.

THE QUANTIFICATION OF VOICE

Seeing sound technologies such as the LRAD as powerful and possibly lethal ways to move bodies in the policing of space is a first step in understanding how sound can demarcate territory. This is particularly apparent if the whole range of sound technologies, including those that rely on voice, are explored. In 2015, *Wired,* the leading tech and economics magazine, published an article claiming that "voice control will force an overhaul of the whole Internet."[10] The author was not being hyperbolic. Over the past decade, the rise of voice technologies and speech recognition across all arenas of communications has been meteoric, from mobile devices, watches, and computers to technologies used in cars, health care, the military, air traffic control, automatic translation, robotics, and aerospace. Speech recognition has also been significantly deployed for purposes of productivity, security, and identification – for the governance of bodies in warehouses, in prisons, and policing and along border zones.

The use of voice recognition in forensics, especially as evidence, has been contentious, mainly because of the issues attached to ensuring full accuracy in the identification of individuals.[11] Many variables must be accounted for such as individual vocal variability (voice changes caused by emotional state, physical state, age, and so on) and recording conditions (compression and background noise). Despite this, voice recognition and identification have been standardized throughout contemporary law enforcement.

The use of voice in determining geographical mobility by nation-states shows just how deeply embedded racial and ethnic profiling is in biometrics. More than fingerprints or irises, voices announce the socialized and encultured characteristics of the speaker. This is nowhere more evident than in the forensic audio systems used on asylum seekers and refugees to impede movement outside of detention centres. In the United Kingdom, between the Asylum and Immigration (Treatment of Claimants, etc.) Act 2004 and the Immigration, Asylum and Nationality Act 2006, compliance with electronic tagging and monitoring transitioned from a fully consensual to a conditional requirement.[12] Although making electronic tracking compulsory would have constituted a criminal abuse (though cases can be made to implement compulsory consent), encouraging compliance as beneficial to gaining asylum has meant that electronic monitoring has been assimilated into the suite of methods determining, and often severely and unjustifiably constraining, the everyday mobilities of asylum seekers. Corporate service providers operating across the prison-detention industry, such as G4S and Serco, are developing more accurate voice-verification systems capable of accommodating multiple dialects, accent variations, and voice and condition changes.

Working in conjunction with the United Kingdom Border Agency, such systems sit alongside security protocols designed to confirm, through accent and dialect, asylum seekers' places of origin. Amid controversy around the nature of language and accent acquisition, the United Kingdom in 2003 ratified the use of forensic linguistics to examine vocal features relevant to the individual's geographical and social origin. The use of language testing by border agencies was, from its inception, criticized by human-rights lawyers and linguists who argued that the assumption that language can be equated to nationality is problematic because it is based on an essentialized model of nationality. Languages and dialects have permeable borders; they change over generations. People who grow up in several areas often have mixed accents or lose their "mother tongue." Language tests also heavily depend on the expertise of the translator.[13]

According to a 2011 report prepared for the United Kingdom Border Agency, in cases of doubt, language analysis was taken up to assist in establishing whether an asylum applicant was from their stated country of origin. UKBA claimed that asylum seekers were falsely identifying themselves as originating from countries that were prioritised for humanitarian protection.[14]

During the pilot program, the United Kingdom Border Agency focused more closely on applicants from Afghanistan, Eritrea, Kuwait, Palestine, and Somalia, countries for which removal and return agreements were available. In 2013, claimants from Syria were added.[15] Although not compulsory, as it is in electronic monitoring, an asylum seeker's refusal to participate in testing may have a detrimental effect on the case. The profiling inherent to these systems of linguistic analysis supports unequivocal racial and ethnic discriminations, further illuminating the already uneven, racist, and ethnically biased nature of the asylum processes.

The prominence of sound and voice in policing and industry applications must be considered seriously. How is data collected? To what ends are they being put? Given that these technologies, for the most part, are invisible and that their effects tend to be unreported or obfuscated, it is critical to simply flag their existence.

We have long debated and discussed contemporary forms of spatial division, the shifting of nation-state boundaries, and the ubiquitous use of computing to determine the movement and flows of bodies, commodities, and services. Less has been said about the place of voice and sound within these processes. The time lapse between the implementation of new technologies and their critique by commentators is substantial, and more needs to be done to investigate the lived effects of voice and sound surveillance, particularly in the labour force and along border zones. It is only through extensive investigation that responses to, and subversion of, new forms of governance can arise. This is especially true as we face a future where shortages of land, work, and resources and growing economic and political unrest will make the free movement of populations and struggles for justice all the more acute.

NOTES

1   Steven Goodman, *Sonic Warfare: Sound, Affect and the Ecology of Fear* (Cambridge: MIT Press, 2009), xiv.
2   R. Murray Schafer, *The Tuning of the World* (New York: Knopf, 1977).
3   STAR-TIDES, "LRAD Corporation – LRAD 100XTM Handheld Acoustic System," http://star-tides.net/lrad-corporation-lrad-100xtm-handheld-acoustic-system.

4   Jürgen Altmann, *Millimetre Waves, Lasers, Acoustics for Non-lethal Weapons? Physics Analyses and Inferences* (Osnabrück: Deutsche Stiftung Friedensforschung, 2008).

5   Ibid., 44.

6   Deborah Cowen, *The Deadly Life of Logistics: Mapping Violence in Global Trade* (Minneapolis: University of Minnesota Press, 2014).

7   Mark Thompson, 2009. "Is There a Sound Defense against the Somali Pirates?," *Time*, November 19, 2009.

8   Juliette Volcler, *Extremely Loud: Sound as a Weapon,* translated by Carol Volk (New York: New Press, 2013).

9   Altmann, *Millimetre Waves.*

10 Cade Metz, "Voice Control Will Force an Overhaul of the Whole Internet," *Wired,* March 24, 2015.

11 Harry Hollien, Ruth Huntley Bahr, and James D. Harnsberger, "Issues in Forensic Voice," *Journal of Voice* 28, 2 (2014): 170–84; Frank Horvath, Jamie McCloughan, Dan Weatherman, and Stanley Slowik, "The Accuracy of Auditors' and Layered Voice Analysis (LVA) Operators' Judgments of Truth and Deception during Police Questioning," *Journal of Forensic Sciences* 58, 2 (2013): 385–92; and Phil Rose, "Technical Forensic Speaker Recognition: Evaluation, Types and Testing of Evidence" *Computer Speech and Language* 20, 2 (2006): 159–91.

12 Joint Committee on Human Rights, *Thirteenth Report,* 2004, http://www.publications.parliament.uk/pa/jt200304/jtselect/jtrights/102/10204.htm.

13 Melanie Griffiths, "'Establishing Your True Identity': The Negotiation of Discourses of Identification by Detained Asylum Seekers in Oxfordshire," in *People, Papers, Practices: Identification and Registration Practices in Transnational Perspective,* ed. Ilsen About, James Brown, and Gayle Lonergan (Hampshire: Palgrave Macmillan, 2013), 281–392. See also Diana Eades, "Applied Linguistics and Language Analysis in Asylum Seeker Cases," *Applied Linguistics* 26, 4 (2005): 503–26.

14 United Kingdom Border Agency, "Language Analysis Testing of Asylum Applicants: Impacts and Economic Costs and Benefits," 2011, https://www.gov.uk/government/uploads/system/uploads/attachment_data/file/257177/language-analysis.pdf.

15 Mark Harper, "Language Analysis Testing Authorisation 2013: Palestinian, Syrian and Kuwaiti (Testing)," United Kingdom, written statement to Parliament, February 25, 2013, https://www.gov.uk/government/speeches/language-analysis-testing-authorisation-2013-palestinian-syrian-kuwaiti-no-2.

Big Data Meet Location Monitoring

James Kilgore

IN THE COURSE of a year, an estimated two hundred thousand people spend time on an electronic monitor following an encounter with the criminal justice system. From May 2009 to May 2010, I was one of them. I suppose I can't complain. Certainly, being on the ankle monitor and living with my family was much better than the six and a half years I spent before that in prison. But comparing electronic monitoring (EM) to prison misses the point.

I have written extensively on the ways in which EM constitutes a form of virtual incarceration that is largely unrecognized by authorities. But there is another aspect to EM that may be even more problematic – how it fits into the notion of the surveillance state.

The first generation of ankle monitors, built on radio-frequency technology, simply informed the authorities whether the wearer was at home. The advent of GPS in the 1990s changed all that. Nowadays, most devices track the wearer's movement in real time. Hanging out in a drug-dealing area or visiting a former partner in crime will show up on a parole officer's screen, perhaps precipitating a return trip behind the razor wire.

In Europe, there has been great debate among probation officers about the human-rights implications of this kind of invasive technology. Countries such as Sweden and the United Kingdom have avoided moving to GPS precisely because they view the location-tracking capacity as excessively invasive. The European Union has even adopted recommendations for EM that posit rights for those on a monitor and cautions against rules that impose unnecessary restrictions on wearers or even their housemates. Germany has taken such cautions to another level, mandating deletion of all tracking data within two months of it being recorded and severely limiting the capacity of law enforcement to use such data in criminal investigations.

By contrast, practitioners, policy makers, and service providers in the United States are only beginning to open the door to a discussion of the rights of the person on the monitor. The vast majority of research focuses on whether EM reduces recidivism or enables crime. Service providers boast of the power of their devices to track and keep vast databases. In contrast to the German approach, one of the major US providers, Satellite Tracking of People (STOP), announces on its website that all tracking data are kept for a minimum of seven years.

If we try to peek into future uses of this technology, things quickly move into the realm of the terrifying. At present, many monitors are programmed with exclusion zones. If a person enters a prohibited area, the ankle monitor triggers an alarm that alerts authorities. Exclusion zones have mostly been applied to those with sex-offence convictions or with a history of gang-related activity. The application of these zones to people with sex-offence convictions has been paired with a number of state laws and ordinances that bar these people from going within a specified distance of areas deemed potential crime sites. In the case of people convicted of sex offences, regardless of the nature of their crime, the prohibited areas are typically areas where children can be found – schools, daycare centres, parks and, in some instances, even bus stops, libraries, and churches. These geographical restrictions also combine with sex-offence registries that post names, addresses, and other information. Every state now has a sex offender list.

Predictably, exclusion zones have prompted intense political struggles. Perhaps the most well known took place in Miami, where a five-hundred-foot perimeter from parks, schools, and so on virtually converted the entire city into an exclusion zone, leaving those on the hot list with no place to live or work legally. When a group of people with sex-offence convictions settled under a bridge, local residents complained. The city then declared a small patch of nearby municipal land a public park, thereby excluding the residents of the encampment from their new-found home. Fed up with constant harassment, a cohort of these people, under the stewardship of Rev. Dick Witherow, packed their bags and moved some ninety miles out of Miami to the rural town of Pahokee, which came to be known as Miracle Village.

In his 2009 book, *The Modern Day Leper*, Dick Witherow character-izes those with sex offense convictions as people whom no one wants and whom few people will defend. In a frenzy of fear and an irrational urge to protect children against an apparent consummate evil, technology-based policies of social control have been passed with no scrutiny of the long-term implications. Exclusion zones have combined with three important social processes to pose a frightening scenario of virtual gentri-fication in our urban areas.

First, we have the growing capacity and presence of monitoring technology. We are all virtually under the monitor and experiencing the expansion of big data, or what Frank Pasquale, in his 2015 book, calls the black box society. All information recorded about us ends up in state and corporate databases that are increasingly mobilized to influence our consumer habits, predict our behaviour, and protect the public against us should we display any disturbing patterns according to the latest al-gorithmic formula.

Second, we have the class- and race-based gentrification of our urban areas. Although the market and urban-development policies are the pri-mary drivers of these changes, policing has also been a huge part of cleansing targeted areas of the visible presence of the poor. As Katherine Beckett and Steve Herbert demonstrate for Seattle in their book, *Ban-ished: The New Social Control in Urban America* (2010), the enforcement of ordinances and the use of petty arrests have effectively banished the poor from the city's privileged urban landscape. Gourmet ghetto super-consumers are free to do their Iyengar yoga classes, guzzle their favourite microbrew, and cruise the streets in their Range Rovers in peace. Ordin-ances, combined with a plethora of fines, have forced exclusion, and often extreme poverty, on the urban unwanted. They do so by banning survival-linked activities such as sleeping in parks, panhandling, or even serving food. Predictably, in 2013, authorities in Silicon Valley added another prohibition to this mix: a law against living in a car.

Third, and intimately linked to the second, we have growing but somewhat contradictory disdain for both the poor and mass incarcera-tion. Across the political spectrum, spokespeople are arguing that we need to reduce incarceration numbers and corrections spending. But most of these cries for action say little about the underlying poverty, inequality, and punitive mentality that is driving what researcher Martin

Horn calls the prison binge. The urge to banish the poor remains. The quest is to find a cheaper way to exclude.

A more efficient alternative might be a form of virtual incarceration that incorporates location monitoring and exclusion zones not only for people with sex-offence convictions or gang histories but also for the full range of undesirables or those adjudged high-risk by algorithms. The widening of the net might lead to the capture of a broader "Google group" – those with a criminal background along with those with histories of mental illness, substance abuse, school suspension, "excessive" reliance on public benefits, conflicts with family services agencies, or extreme levels of debt.

This linkage of databases with location (through cellphone monitors, chip implants, or whatever the future holds in store) is a variation on one of the dominant themes in science fiction: keeping certain people in and certain people out without having to mobilize troops, build walls, or continue the prison-building mania. As a sequel to mass incarceration, mass monitoring has the potential to satisfy those with a penchant for adopting quick technological fixes for complex social problems. It could be a way to keep global cities at the cutting edge of world-class lifestyles for those kept in and a way of promoting mass immiseration for those kept out.

Digital Apartheid

Visualizing Impact

VISUALIZING IMPACT IS a nonprofit collective that creates data-led visuals for social justice. Visualizing Palestine, our first and largest project, promotes a factual, rights-based narrative of the Palestinian-Israeli situation.

Between 2012 and 2019, Visualizing Palestine published more than one hundred infographics under a Creative Commons license and two web platforms centred on human rights data and information. Users download these visuals on a daily basis to engage in advocacy, education, and cultural activities. As of 2018, Visualizing Palestine users came from at least four hundred cities and sixty-five countries. The following infographics on Palestinian digital connectivity illustrate the results of Israeli control over the Palestinian telecom sector: a digital divide between Israelis and Palestinians. Israel controls Palestinian access to the electromagnetic spectrum, coverage areas, equipment imports, infrastructure installation and maintenance, and international gateways.

HOW LONG TO DOWNLOAD A TV SHOW?
ISRAEL DECIDES.

Israel controls and restricts the development of the Palestinian telecom sector, leading to slow internet speeds in the West Bank and Gaza.

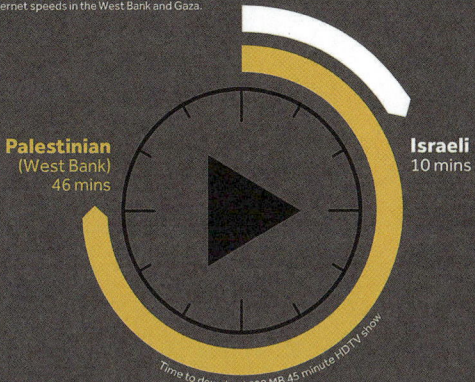

Palestinian
(West Bank)
46 mins

Israeli
10 mins

Time to download 600 MB 45 minute HDTV show

OCCUPIED AIRWAVES
PALESTINIAN DIGITAL CONNECTIVITY

3G Wireless
Israel controls access to wireless frequencies. Palestinian operators prevented from providing 3G in the West Bank until 2018. Still no 3G in Gaza.

TV and Radio
Israel controls the radio frequency spectrum. Limited frequencies available for Palestinian stations.

Broadband Internet
Israel controls development of infrastructure. Just 38% of Palestinian households have connectivity above 4 Mbps, versus 94% of Israelis.

PAYING MORE FOR LESS
BROADBAND PRICES FOR ISRAELIS AND PALESTINIANS

Israel's restrictions on the development of the Palestinian telecom sector result in Palestinians in the West Bank and Gaza paying higher prices for slower internet speeds.

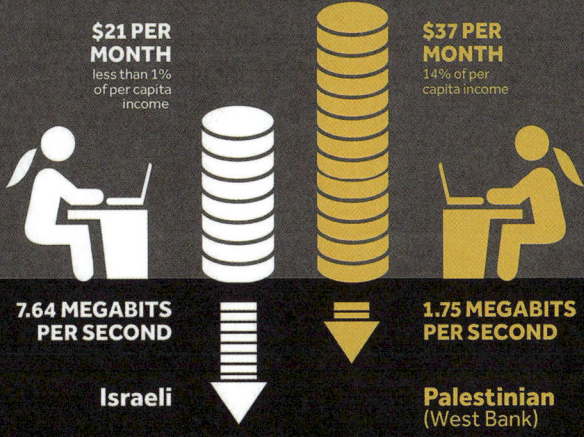

$21 PER MONTH
less than 1% of per capita income

$37 PER MONTH
14% of per capita income

7.64 MEGABITS PER SECOND

1.75 MEGABITS PER SECOND

Israeli

Palestinian
(West Bank)

Campa Cola is built.

Developers illegally seize land from tribal communities on the outskirts of Bombay, facilitated by mafia violence and government corruption.

Mira Road municipality is formed.

1967   1971   1976   1980   1985   1989   1991

First Development Plan for Greater Bombay & Regional Planning Authority are created.

Plans include decentralizing urban population and industry to the suburbs.

Maharashtra Slum Areas Act created to declare a "slum area," and clear and redevelop it.

Urban Land Ceiling Act limits land ownership in Mumbai. Development shifts to the edge of the city.

Slum Redevelopment scheme (SRD) introduced. It allows developers to demolish slums and build new on-site housing, if 75% of residents consent.

The globalization of real estate begins. Mortgage-backed securities become popular.

IMF and World Bank policies create a more market-driven economy.

MUMBAI

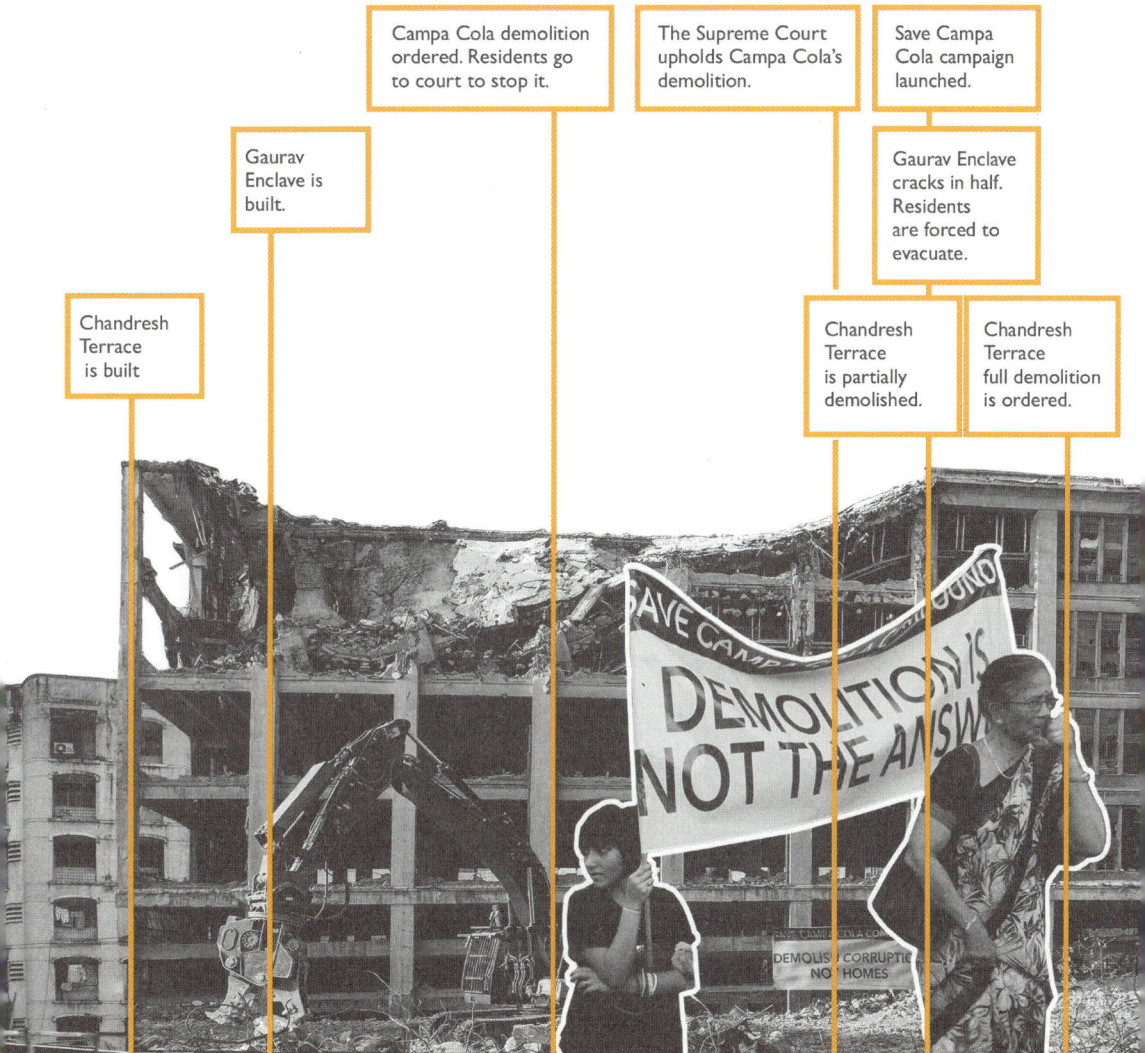

Campa Cola demolition ordered. Residents go to court to stop it.

The Supreme Court upholds Campa Cola's demolition.

Save Campa Cola campaign launched.

Gaurav Enclave is built.

Gaurav Enclave cracks in half. Residents are forced to evacuate.

Chandresh Terrace is built

Chandresh Terrace is partially demolished.

Chandresh Terrace full demolition is ordered.

1996 1997   1999 2000   2005   2007   2011   2013   2015 2016

Slum Rehabilitation Authority created. It is the single point of contact and decision-making for all slum redevelopment projects in Greater Mumbai.

Jawaharlal Nehru National Urban Renewal Mission dismantles rent controls and relaxes zoning laws.

The Special Economic Zone Act expands state powers of eminent domain.

Urban Land Ceiling Act repealed.

The Real Estate (Development & Regulation) Act protects home-buyers and regulates the industry.

Real Estate Investment Trusts become popular.

Foreign investment in real estate rises.

Restrictions lifted on foreign direct investment in real estate. Large real estate and investment firms set up shop in India.

The mortgage foreclosure epidemic becomes a global financial crisis.

International corporations invest almost $1B in Indian real estate firms, even as the sector slows.

MUMBAI

Mumbai Rising, Buildings Falling

Emily Paradis, Brett Story, Deborah Cowen

MUMBAI IS INDIA'S budding international financial centre.[1] It is one of a growing crop of cities in the global south remaking themselves as "world-class" in the international network of finance capital. The digital circulation of investment capital, particularly into city building, has been integral to restructuring the global urban political economy. Although real estate has historically been viewed as a local issue, the last three decades have seen the rise of a global market. The range of actors that shape the real estate industry – from builders and brokerage firms to consultants, finance companies, and investors – "have all extended their area of operations beyond local markets to a world-wide base."[2] India is at the centre of these transitions. As Saskia Sassen explains, "Home mortgages today are a new frontier for using high-risk financial innovations to extract profit," and she flags India as one of the global south countries leading the way in the growth of mortgage markets.[3] It is not simple finance in Mumbai but the financialization of Mumbai that has unfolded. Yet while Mumbai is rising, its buildings keep falling.

Across the city, but most intensively in the rapidly developing outer rings, high-rise buildings, many if not most of them occupied, are being demolished or slated for demolition at an unprecedented pace. The pretexts for this demolition range from dangerous structural deficits to juridical irregularities that deem the buildings "illegal." Between 2008 and 2013, more than 55,000 illegal buildings were counted in Mumbai.[4] In the northern suburb of Mira Road, it is estimated that more than 80 percent of buildings are illegal. Dangerous and illegal buildings are not new in Mumbai, but they have suddenly been charged with new meaning in an urban drama that has ushered in a wave of displacement, across class and housing tenure.

The ascendancy of Mumbai's economy and the demolition of its buildings are, in fact, two sides of the same coin of urban real estate speculation. Through the lens of three sites, we trace the co-constitution of these twinned phenomena, the expulsions they enable, and the hybrid digital-material forms of resistance that communities deploy and develop in defence of their homes.[5] Residents' circumstances and their claims reveal the contours of an emerging advanced capitalist economy of re-development in which the very category "illegal buildings" has proven to be not a detriment to building developers but rather a boon. It presents a legal means to circumvent one of the most sacrosanct pillars of liberal capitalism: property rights. Financialization is both creating this build-ing frenzy and providing the conditions and logics for the destruction of existing structures, and it is utterly reliant on digital technologies. Here, we tell a complex, somewhat concealed story of digital life that may at first glance not seem particularly digital at all. The story has less to do with devices and technologies and more to do with the digital as an infra-structure for particular forms of urbanization and their contestation.

THREE STORIES OF (RE)DEVELOPMENT

Our research team made multiple visits to Mumbai between 2012 and 2014 to examine how three developments – transformations in the city's real estate economy, the crisis of illegal buildings, and emerging digital technologies – had intersected to reshape everyday life and practices of urban citizenship, particularly in the suburbs. We came to know three communities in particular, communities that pointed us towards illegal buildings as an important terrain of struggle, a terrain on which the circulation of global capital and digital communication technologies are bearing concretely on the lives and buildings that occupy Mumbai's outer rings.

One of these communities is in the Worli area at the heart of South Mumbai while two are in Mira Road, a northern suburb. Worli – origin-ally one of seven islands joined together by landfill in the eighteenth century to create the city of Bombay – was an industrial zone of manu-facturing plants and textile mills until the 1980s, when the advent of synthetics, the demands of striking textile workers, and India's economic liberalization transformed Bombay from an industrial city in which three-quarters of workers were organized to a neoliberal global city in which at least two-thirds of all jobs are in the informal sector.[6] Worli became a microcosm of this economic shift, its mill lands redeveloped as high-rise flats for the city's emerging middle class.

Campa Cola, our first site, is a product of Worli's early redevelopment. The compound's seven towers were built in the 1980s on the site of a former soft-drink plant. They were zoned to be five storeys each, but the developers exceeded the permitted height on several buildings, in one case by sixteen floors. In the words of one resident leader, Worli was "not posh then." Modest flats in a deindustrializing neighbourhood were purchased without occupancy certificates by young middle-class families, many of whom spent their life savings on their apartments and raised children to adulthood in them.[7] Residents' early years in the compound were spent in a battle for basic services such as running water, which the city long refused to provide because of the buildings' legal and structural irregularities.

Though access to city services was eventually granted, the dispute led to a determination by the Brihanmumbai Municipal Corporation, Mumbai's municipal government, that the building's unauthorized storeys were illegal. Since 2005, Campa Cola's residents have been fighting courts and demolition crews to prevent the city from razing the buildings (see "Dispatch from Mumbai" in this section for images and more). Now a posh area, Worli is the site of feverish speculation on land and density allowances for superluxury developments – such as World One, billed as the world's tallest residential tower – marketed to international elites and wealthy Indians.[8] A development application on an adjacent parcel of land still occupied by the defunct cola factory can't move forward because the Campa Cola buildings have consumed the allowable density for the whole plot.

Mira Road, home to our other two sites, emerged from the same development boom that created Campa Cola, whose reverberation into peri-urban Mumbai was fuelled by India's Urban Land Ceiling (Regulation) Act and regional planning initiatives to promote population dispersal to the suburbs.[9] Just outside Mumbai's northern boundary, five villages governed by *gram panchayats* (village councils), merged to form the Mira-Bhayander Municipal Council (MBMC) in 1985. The municipality has grown rapidly in the intervening thirty years, adding several other villages to its territory and transforming from a protected saline estuary and agricultural zone into a dynamic city with an estimated population of 1 million or more. Builders worked with municipal officials to seize land for development from farmers and tribal communities through fraud and violence.[10]

Today, in addition to the remaining tribal Agri and Koli inhabitants and Maharashtrian villagers, Mira Road's residents include diverse groups: in-migrants drawn to Mumbai's economy from the surrounding rural areas of Maharashtra and all over India; first-time home buyers from central Mumbai seeking a more affordable housing market; a large Muslim community, including those violently displaced by the 1992–93 Mumbai riots; a Nigerian community that settled on land granted by the British colonial government for military service; undocumented workers from Bangladesh who labour, and sometimes live, on the city's sprawling construction sites; and communities moved from informal settlements in central Mumbai into private buildings on the periphery through various state-sponsored slum-redevelopment schemes.

Though the MBMC became a municipal corporation in 2002, Mira Road still lacks some of the amenities and structures normally in place for neighbourhoods in large cities as well as public infrastructure such as hospitals and schools. As is the case in other parts of Mumbai, Mira Road is run by Mafia, which operate openly and form another tier of governance with power equal to, or greater than, that of the municipal government.[11] As a result, the vast majority of buildings that house Mira Road's ever-expanding population have been constructed with little regulation and oversight.[12] Many are built on unsuitable sites with substandard materials, particularly concrete made with water and sand dredged from the estuary. The concrete's high saline content causes it to degrade rapidly. As was the case in Worli in the 1980s, new buildings in Mira Road generally have no municipal services. This lack of services serves the interests of the water Mafia, which sells tanks of water – often obtained illegally from reservoirs intended to irrigate nearby farms – at inflated rates to building residents.

Gaurav Enclave, the second building or site, does not have adequate water infrastructure but its digital infrastructures are more pervasive. This massive 234-unit complex was built in 1999 on newly drained salt flats at the city's edge by a company whose buildings routinely exceeded height and density permits.[13] Profiled in a marketing video by 99Acres as "luxury at affordable prices," the flats were snapped up by middle-income workers, most of them first-time owners who had previously

Gaurav Enclave. Photograph by Anushree Fadnavis. Courtesy of the National Film Board of Canada.

rented in Mumbai but could never afford to buy there. Many continued to make the harrowing daily commute to downtown Mumbai, more than an hour each way on packed trains.[14] When residents took possession of their units (like Campa Cola, without occupancy certificates), there was no paved road, no sewage or drainage systems, and no connection to municipal water. The proud new flat owners had to wade through mud to get to their building's front doors. The first order of business for the newly formed Building Society was to have a septic tank dug in the sprawling complex's interior courtyard; it would be more than ten years before the city would install pipes to replace the area's flood-prone roadside sewage ditches.

By 2005, the building had already begun to deteriorate. It had developed a large exterior crack that the Building Society had repaired. In 2013, it suddenly cracked from top to bottom, splitting the wide building down the middle. At about the same time, a building collapse in nearby Mumbra claimed dozens of lives. In the wake of that catastrophe and the media furor about dangerous and illegal buildings that followed, the MBMC moved uncharacteristically quickly to secure residents' safety, ordering residents to evacuate Gaurav Enclave.

By contrast, in our third site, the MBMC made the building hazardous for its occupants. Chandresh Terrace was built in 1996 in central Mira Road, adjacent to a lively commercial street known as "the India-Pakistan border" because it divides the city's two original neighbourhoods, one predominantly Hindu, the other Muslim. Most residents of the compound's four small buildings purchased their modest flats from the original owners. The occupants are a multifaith group of lower-income families; many work close by in Mira Road's service sector or as informal vendors, and some supplement their household finances with temporary jobs in the oil economies across the Arabian Sea.

By 2011, their apartments looked run down – in part because of the Building Society's failure to make needed repairs and upgrades. But an audit showed the complex to be in acceptable condition. For this reason, the arrival of a city-ordered demolition crew one morning in 2013 came as a shock to many. One resident was in the bath when a thunderous impact shook her wall; most others were away at work and school. By the time the incredulous flat owners obtained the documents needed to halt the demolition, the compound's "A" Wing had been partially destroyed.

◄ In 2013, Gaurav Enclave suddenly developed a crack from top to bottom. Photograph by Anushree Fadnavis. Courtesy of the National Film Board of Canada.

▲ The partially demolished "A" Wing of Chandresh Terrace. Photograph by Anushree Fadnavis. Courtesy of the National Film Board of Canada.

▼ Chandresh Terrace rubble. Photograph by Anushree Fadnavis. Courtesy of the National Film Board of Canada.

Residents later obtained through a right-to-information request a document that revealed that members of the Building Society had colluded with a developer who hoped to profit from an anticipated municipal plan that would multiply the site's density allowance.[15] By submitting a fake structural audit and photos of a different location, they had Chandresh Terrace added to the list of dangerous buildings being rapidly compiled by the MBMC in the aftermath of the Mumbra catastrophe. The community was torn apart after the revelation, some residents taking up the builder's offer to relocate while redevelopment proceeded, others defiantly staying on in the rubble to protect their ownership rights.

TRACKING DIGITAL RESISTANCE

At Chandresh Terrace, Gaurav Enclave, and even posh Campa Cola, the optics of residents' struggles – women berating demolition crews, youth facing off against police, residents being displaced and dispersed, families defiantly refusing to leave homes reduced to mud and rubble – evoke long-familiar images of campaigns to halt slum clearance. These images circulated not only in local neighbourhoods but also in transnational mainstream and social media. But residents' organizing in response to the crises in their buildings also took new forms, enabled by information and communications technologies.

The significance of digital resistance was most apparent at Campa Cola, where residents fought for their homes in the courts, on the streets, and online. In the wake of an unfavourable court decision that set a November 2013 deadline for their eviction, a group of young people who grew up in the compound launched a campaign called Save Campa Cola.[16] The campaign made use of sophisticated design and messaging, traditional lobbying and media relations, and PR actions such as a hunger strike and a multidenominational Prayer Week, all amplified through Facebook, WhatsApp, YouTube, and Twitter. It went viral. For months, blogs and front pages across India carried news of the impending eviction, the arrival of bulldozers and police on eviction day, the residents' refusal to leave, the temporary reprieve they were granted, and ongoing wrangling in the courts. In the lead-up to the 2014 federal election, candidates tweeted their sympathies and promised intervention. The case became a flashpoint for debates about the builder-authority nexus and the plight of the middle class.[17]

▲ Some Chandresh Terrace families remain in their flats to keep the building from being completely demolished. Photograph by Paramita Nath. Courtesy of the National Film Board of Canada.

▼ A view through the walls of a Chandresh Terrace flat. Photograph by Anushree Fadnavis. Courtesy of the National Film Board of Canada.

Digital technologies also figured prominently in residents' struggles at Mira Road. Following the appearance of the crack at Gaurav Enclave, the building society started a Facebook page as a way to communicate and organize. The public page alerted our research team to their existence and offered a poignant chronology: residents' expressions of disbelief and dread in the days following the appearance of the crack; the evacuation order issued by the city; the dispersal of the community into rented apartments across Mira Road; their pleas for financial compensation; heated debates about whether to work with the builders on redevelopment or take them to court; painstaking compilation of the signatures required to embark on a redevelopment plan; and reams of official letters from the MBMC and the builders. Then, finally, after a year-long virtual silence, followed a triad of auspicious postings: photos of a *bhoomi pooja* (a performance) celebrated on the site of the planned new building; an artist's rendering of a shiny new tower (a tall, airy contrast to the first building's outdated concrete bulk); and a letter from the builders thanking residents for their "sacrifice and patience" and announcing a limited-time offer for owners to purchase additional space for their new flats at a discounted price – thereby "transforming your lifestyle from old to new homes."[18]

A young Chandresh Terrace resident records meetings with corrupt officials on her phone.
Photograph by Anushree Fadnavis. Courtesy of the National Film Board of Canada.

Even in the lower-income community of Chandresh Terrace, where mobile phones are ubiquitous but internet connectivity is scarce, digital technologies played a pivotal role. Residents who had no prior relationships with each other began sharing large portions of their day online. One young resident who became active in organizing for the building's future reported that she and the other active residents became virtually glued to one another through their phones and WhatsApp. A young resident of "A" Wing covertly recorded her meetings with officials on her phone, delivering audio proof of the fraud that had led to the demolition of her home.[19] And it was by tracing the MBMC's online procurements for demolition that the residents proved that the demolition had taken place before the competition for the demolition contract had even closed. The story of the remaining families squatting in their demolished homes – and especially of one teenage woman's organizing efforts to protect the apartments – was picked up on digital news media, which circulated all the way to Toronto.

Through research and documentary work with residents, scholars, activists, and journalists, we aimed to situate these stories, and the broader crisis of dangerous and illegal buildings of which they were a part, within the context of the feverish speculation and redevelopment driving Mumbai's emergence as a global city. We sought to document the new technologies of dispossession that were being developed and deployed, seemingly before our eyes, as well as the nascent forms and constellations of resistance that were emerging in response.

PROPERTY IN QUESTION

While there is much to set Campa Cola apart from the buildings in Mira Road, especially the discrepancies in wealth and class, in each of these cases the mutable categories of "legal" and "illegal" – as applied to construction, occupancy, and tenure – have thrust residents into new forms of precarity. And though flat ownership is in many respects a world apart from life in informal settlements, it is possible to draw a line from slum clearance to the crisis of illegal buildings: some of the tactics and legal mechanisms that have long been employed to undermine slum dwellers' claims and rob them of their homes are now increasingly wielded against building residents.[20]

Once home to a thriving rental sector, Mumbai's primary tenure is now ownership, even among lower-income households.[21] Living "on rent" is an option that even very low-income households will consider only as a temporary last resort. With its promise of wealth accumulation, security of tenure, and permanence, ownership is pursued at all costs – but its shifting legal boundaries increasingly place purchasers in the same precarious position they seek to leave behind, or even worse off because of crippling debt.

Debt is indeed a key contributor to this precarity. For low- and middle-income households in Mumbai's polarizing labour market, purchasing a flat introduces mechanisms of debt – bank loans, mortgages, building-society fees – that facilitate dispossession. At Chandresh Terrace, for example, some residents went along with the builder's fraudulent scheme because the company claimed it would pay off their outstanding building-society fees, loans, and arrears for gas and water services. Gaurav Enclave residents had to continue to pay their mortgages while displaced, a cost that doubled when the builder failed to compensate them for their temporary rental accommodation.

The residents' struggle to make ends meet made them prey for speculators who bought their flats – or more accurately, their rights to whatever new flats might one day materialize – at half their value, leaving the sellers to pay off a mortgage on an apartment they no longer owned or inhabited.[22] Ownership became a form of indenture, harnessing them to debt that remained long after their property was gone. The flats might generate wealth – but that wealth accrues to developers, creditors, and speculators, not owners. This pattern has eerie similarities to the subprime mortgage crisis in the United States except that, as an informant pointed out, they were not "bad debts": few people ever default.

Ownership of the flats is not simply financially precarious; it is insecure because ownership rights are attached to the unit and building, not the land. Once a building is demolished, these rights come into question, particularly if owners lack legal documentation such as occupancy certificates. Families stayed on in the rubble of "A" Wing at Chandresh Terrace, for example, because they were afraid they would lose their claim to the property if the demolition was completed. Gaurav Enclave residents were desperate to reach a legal settlement with the builder for the same reason, even if it meant compromising their entitlement to compensation. This form of ownership is the very legal mechanism that provides an incentive to raze informal settlements. As we learned from activists with Youth for Unity and Voluntary Action, it is also the reason that the land-based slum resettlement programs of the 1970s and 1980s – in which displaced households received deeds to land on which to build new homes – provided more durable rights than the building-based resettlement schemes that have been in place since the 1990s.

If the unit and building are the owners' only real property, unauthorized construction and redevelopment can render these possessions materially and legally impermanent. Thousands of buildings in Mumbai and Mira Road have been developed illegally as municipal and state governments have failed to provide regulation, oversight, or enforcement.[23] Gaurav Enclave, hastily constructed on marshy land with poor-quality materials that degrade within a few years, is just one of many dangerous and deteriorating buildings produced by this laissez-faire regime.

But even when apartments are structurally sound, legal irregularities can provide a pretext for selective enforcement and demolition, as was the case for Campa Cola. Multiplying land values, new density allowances, and rising housing demand can also fuel schemes to redevelop buildings long before the mandated thirty-year period has ended, as happened with Chandresh Terrace. The same tactics of illegalization, fraud, threat, and violence – long used to displace and dispossess farmers, tribal peoples, slum dwellers, and inner-city tenants – are now being applied to suburban flat owners who stand in the way of redevelopment.

Though one might expect that residents' formal ownership of their units would provide them with additional protection not enjoyed by these other groups, the opposite is true, at least where legal recourse is concerned. Although farmers, tribal peoples, slum dwellers, and tenants are, at least nominally, protected by national and international rights regimes, flat owners have no collective protections.[24] Their claims are instead mediated by a market regime of contractual, privatized rights and risks. Their cases are governed by corporate and contract law, which treat homes as private property, not by rights regimes that recognize homes as a social need and see residents as being owed a duty by the state.

For flat owners such as those at Gaurav Enclave and Chandresh Terrace, most of whom had no prior knowledge of legal and regulatory procedures, it is unclear whom they should hold responsible when their homes are destroyed. The state? the builder? one another? themselves? This uncertainty leads to a confusion of courts and claims, and even lawyers acknowledge there is no definitive way to proceed. In the case of Chandresh Terrace, for example, multiple civil and criminal proceedings were launched. Some were brought forward by individuals, others by collective entities; each had different ramifications for the other proceedings and for the future of the compound. Resolving them might take years while residents are left in legal limbo, watching the bulldozed walls of "A" Wing come closer to collapse with each monsoon. The piecemeal legal and regulatory apparatus to which flat owners have access stands in stark contrast to the consolidated system that expedites development. This situation, too, mirrors a discrepancy long enshrined in slum redevelopment. As one report remarks, "There is a single window for the developers but slum residents have to run from pillar to post for getting information, responding to actions of the state, getting their entitlements and to extract some response in the case of adverse impacts."[25]

In the context of planning regimes and market conditions that incentivize unsanctioned (re)development, the occupancy certificate has emerged as a particularly useful technology to privatize and mystify owners' rights and risks. On its face, it is intended to protect purchasers, certifying that their building and unit meet regulatory and safety requirements. In practice, it is widely disregarded: in 2014, there were an estimated six thousand buildings in Mumbai without occupancy certificates, up from two thousand in 2010, and in Mira Road the numbers are far worse.[26] Neither the state nor private entities meaningfully enforce the requirement. Lack of a certificate does not prevent developers from selling units, nor does it keep buyers from obtaining a mortgage from a bank or paying property tax to the municipality. Until recently, the existence of certificates was little known to the general public, and at no point in the transaction are purchasers informed about them. However, the lack of an occupancy certificate can be used to call an owner's property rights into question.

In the cases we examined, questions of responsibility, and therefore liability, turned on the existence, or lack, of an occupancy certificate. In the Campa Cola case, for example, the courts and news media suggested that, in taking possession of their units without certificates, the residents had colluded with the unsanctioned development and were therefore responsible for their own predicament. The occupancy certificate thus functions to shift responsibility for illegal development onto individual flat owners and away from the builder's contravention of regulations and the state's failure of enforcement.

With the media attention surrounding the Campa Cola case and illegal buildings in general, public awareness of certificates has increased. This awareness, however, has not led to a more consistent enforcement regime. Certificates have instead become a new vehicle to sell flats and further polarize the real estate market. Developers now tout the availability of certificates alongside other amenities in their promotional materials. Purchasers who can afford it pay a premium for units that have certificates, while those who can't must purchase units in unsanctioned buildings and bear responsibility for the consequences.

BECOMING MIDDLE-CLASS

The privatization of risk and responsibility is linked to the construction of flat owners as universally middle-class. This construct brings neoliberalism into the home, operating to situate these households outside collective rights, with the attendant expectation that they meet their members' needs without state intervention. Flat ownership itself is understood as synonymous with middle-class status, drawing the precarious service-sector workers and migrant labourers of Chandresh Terrace, the mid-level service workers of Gaurav Enclave, and the transnational professionals and merchants of Campa Cola into a presumed set of shared interests and investments. Despite wide variations in the financial vulnerability of flat owners – including former residents of informal settlements moved into buildings through slum redevelopment schemes – even many NGOs still consider suburban high-rise dwellers to be outside the category of "urban poor," people who might be in need of their advocacy and services.

The built form of the high-rise also reinforces privatization. A number of informants who had moved into the buildings from slums commented on the absence of public space and the isolation of closed doors. With their daily lives structured by the waged labour required to maintain their mortgages, debt servicing, utilities payments, and building fees, residents are often dissociated from one another – until a crisis arises. Chandresh and Gaurav residents, for example, noted that there had been little contact or sense of community among residents until their buildings were threatened with demolition. The fragmentation of a community into self-interested private units leaves residents vulnerable to manipulation and corruption. This is especially the case when, as at Chandresh Terrace, residents have little understanding of administrative processes or when, in the words of a Gaurav Enclave resident, they are "all looking the other way" as they work and commute long hours to make ends meet.

Privatization also leaves residents ill-prepared for the complex collective negotiations, solidarity, and struggle necessary to defend their homes. The vicissitudes of digital-technology access and social-media use can exacerbate the divides that already render cross-class solidarity building so challenging, even as technologies provide new avenues for collective recognition and coalition building. At Campa Cola and Gaurav Enclave, social media offered accessible platforms for rebuilding collectivity and taking action among residents, some of whom had previously felt siloed in their vertically organized flats and disconnected from their neighbours. In the process, residents developed hybrid digital-material forms of resistance. These proved to be both powerful and problematic insofar as access to the technologies and the social connections needed to leverage them politically was unevenly distributed along lines of class and education, among other axes.

This phenomenon bore itself out most explicitly in our observations of Campa Cola residents' savvy social-media use, which, while tremendously useful to their organizing, also reified elitist narratives regarding who fulfills normative ideals of citizenship and, thus, whose displacement constitutes a civic violation worthy of press coverage and political attention. Indeed, Campa Cola's campaign made use of sophisticated conventions of image and message discipline borrowed from corporate and political power centres. Social-media posts and news interviews conveyed images of Campa Cola as a close-knit community composed of cosmopolitan, devout, ordinary, relatable families and seniors who stood to lose everything – as unsophisticated buyers bilked by powerful builders. The campaign warned of the dangers to other self-identified middle-class and ordinary Indians if the demolition was allowed to proceed.

Although successful in capturing media attention, framing the message, and eliciting public sympathy, this simple framing of the issue – aimed at a particular, influential public – was at odds with a nuanced understanding of the economic and political implications of these events and their connections with other similar events in Mumbai, in India, and around the world. Instead of building solidarity against real estate speculation, which was dispossessing ever broader segments of the population, some of the campaign's politically expedient messages instead undermined the claims of less powerful groups – as when Campa Cola residents suggested that they had fewer rights than slum dwellers.

The Gaurav Enclave Facebook page, meanwhile, served as a vital meeting space, especially when residents were dispersed across the city after the building was evacuated. It also enabled the coordination of in-person meetings and allowed for some more public claims-making and media attention. The page became a site for circulating information, airing disagreements, sharing commiseration and support, and negotiating solidarity. These functions likely made the difference in the apparently successful resolution of their crisis.

But the page's public nature also presented disadvantages. For example, in lieu of physical meetings, the builders entered the virtual space and used it to post messages and updates to the dispersed owners. Some residents responded to the builders' posts with strongly worded demands for information and desperate pleas for rent cheques, but these communications appeared to go unanswered, like cries in the virtual wilderness. The conventions of virtual communication and the residents' varying levels of sophistication in the uses of social media shielded the builders from confrontation and diminished the solidarity-building potential of individual residents' well-reasoned, funny, or heart-rending public declarations.

Chandresh Terrace – the most precarious of these communities – in fact made some of the most creative uses of online organizing. They briefly had a Facebook page but found it was not helping their work or resonating with the residents. But they did use information and communications technology to continue their collective organizing and to make claims with evidence against the municipality and building society. Again, WhatsApp became a crucial tool to facilitate virtual meetings and organizing. Digital devices were also critical to exposing municipal fraud, developing more sophisticated claims, and residents building confidence in their organizing abilities.

Though they offer the promise of direct communication across class and geographical lines, these virtual forums don't (yet) transcend the separate struggles they were developed to support. In fact, they risk replicating, and perhaps even strengthening, social and economic divisions in (and the divergent interests of) communities affected by the same macroeconomic forces. As is the case in old-school, off-line social movements, the faction with the most resources gets to define the terrain of debate in its own interests.

In the end, the media spotlight on Campa Cola may have had the contradictory effect of entrenching an image of privileged, middle-class flat owners, thereby obscuring similar dramas unfolding in less prominent neighbourhoods across Mumbai and its suburbs. However, digital technologies are not simply potential platforms to counter power: in many ways, they provide the very infrastructure for property speculation and real estate financialization that these communities are fighting against. If digital technologies can facilitate community and collectivity within and across neighbourhood spaces, they can also produce the city itself as a globalized stock market.

THE CITY AS STOCK MARKET

It is possible to draw a line between slum redevelopment as one type of urban dispossession and the widespread eviction of high-rise dwellers, even those with ownership tenure. No doubt, the figure of the apartment owner looks very different from the slum dweller. Their location within relations of production and consumption, their sense of collective or class identity, their everyday lives and social networks, the organization of the public and private in their lives are all distinct. This is precisely why it is surprising to note a convergence in their experiences of displacement. But looks can be deceiving.

Indeed, despite the differences between these figures and their social locations, they are both subject to a form of state power that dispossesses through the theft of land. In each case, the state is actively involved in dispossessing residents of their homes. In the case of the slums – the residents' tenure is precarious by virtue of their "illegal" settlement status. Slum dwellers, in the interests of development and private capital, can be rounded up by force and literally moved by the state. In the case of high-rises, the residents' vulnerability to displacement does not stem from their illegal occupation but from the building's status. It is the building's illegal status that renders residents precarious. And it is not simply a temporary displacement at stake but a full-scale dispossession of land. The ownership status that supposedly distinguishes apartment dwellers from slum dwellers is profoundly precarious in the face of debt. Apartment owners often struggle to maintain payments on their purchased flats while they secure alternative housing as their buildings are being demolished. Dispossession through debt is a common outcome.

In this context, property rights are rendered precarious through debt and the selective application of regulatory mechanisms. Furthermore, the ideology of ownership constitutes apartment owners as middle-class, with responsibility for bad builders or bad buildings being privatized and falling on the family unit. Although dispossession is increasingly technical and done through market mechanisms, it is no less destructive to people's lives and livelihoods.

How did forms of dispossession that are all too familiar in informal settlements come to extend to private suburban high-rises? What allowed an economy to emerge around the destruction and redevelopment of unsafe and illegal buildings? What kind of economy is being built, literally and figuratively, on the widespread theft of urban land? And what does all this mean for the residents, some of whom worked hard to get from the slums to the towers only to find themselves homeless again but with massive mortgage debt?

To answer these questions, we need to return to the issue of the financialization and internationalization of real estate in Mumbai. The pressure-cooker conditions of real estate in Mumbai have put land values on par with those in New York City. These conditions have given rise to a new form of dispossession, and these conditions were created through specific policies of the Indian state. Once the industrial heart of India, the decline of the city's cotton mills in the 1980s sunk the city into a period of deindustrialization and economic decline. The reconstitution of Mumbai as a globalized financial centre has been built on the city's industrial decline, quite literally. Globally, the city has transformed its image from the industrial capital of India to a world-class finance centre; on the ground, this transformation has meant clearing land for real estate speculation and capital investment.

This process began in the early 1990s. In 1993, the international consulting firm, McKinsey, issued a report lobbying for Mumbai's development into a global financial centre with the goal of turning Mumbai into Shanghai. McKinsey made two key suggestions for change, the "relaxation of land acquisition rules and strict control of labour."[27] In fact, the transformation of the city relied on the literal and figurative removal of working-class people and land uses from specific areas – specifically, Dadar-Parel, the Port Trust lands in the dock area, the Bandra Kurla Complex, and BDD chawls in Worli – and an increase of the floor space index (FSI) in already built-up areas as incentive for redevelopment.[28] By mid-2008, almost the entire mill area had been shaped into a modern commercial hub with global chain stores, corporate houses, and hypermarket complexes.[29] Proactive pursuit of real estate mega projects has been a primary component of the plan.[30]

As Hussain Indorewala and Shweta Wagh note in their contribution to this book, the most recently unveiled development plan has reformed Mumbai's previously "restrictive" FSI system. Increasing the FSI has been identified as central to revitalizing the city's real estate market and economic vitality, so much so that the World Bank has been asserting pressure on the city to increase its FSI allowances for at least a decade. In a city already teeming with people, the only way to increase density is to build up. But before the city can be built up, its existing buildings – even those owned and occupied – must be torn down.

Indian cities have only recently caught the eye of actors in the transnational real estate markets. In "Mortgage Capital and Its Particularities," Saskia Sassen describes how globally linked financial markets have facilitated the deployment of new instruments for the investment of global capital into emerging urban real estate economies, such as Mumbai's.[31] Since the late 1990s, the Indian government has put considerable effort into increasing, systematically, foreign direct investment (FDI) as a percentage of gross domestic product.[32] Today, some of the largest real estate firms in the world such as Morgan Stanley, Blackstone, and Apollo are setting up shop there. This new-found interest can be traced to a piece of Indian legislation enacted in February 2005. Through it, the Finance Ministry relaxed government restrictions on FDI in real estate by lifting the city's 40 percent cap on foreign ownership. The liberalized FDI guidelines opened the floodgates to foreign investment in India's real estate market to such a degree that the real estate index of the Bombay Stock

Exchange doubled over the following ten months. The flow of FDI into Indian real estate quadrupled soon after, causing land values in Indian cities to increase by as much as 500 percent throughout the 2000s. This has led leading economists to describe Indian cities as property-driven stock markets.[33]

Other important policy changes occurred in 2005 that increased capital pressures on Mumbai's real estate market. For example, backed by US$20 million, India launched the Jawaharlal Nehru National Urban Renewal Mission, which put in place a system of incentives to encourage municipal governments to liberalize their land markets through the dismantling of rent controls, the creation of municipal bond markets, the abolition of land-ceiling acts, and the relaxation of strict zoning laws.[34] India's Special Economic Zone Act was also enacted that year. The act expanded the state's powers of eminent domain to create "superliberalized enclaves run ... by private developers."[35]

Perhaps the most significant land dispossessions and property investments are taking place not in India's city centres but rather on the booming peripheries. It is in places such as Mira Road that we are seeing the rapid influx of surplus capital into the emerging real estate market. One consequence of this rapid influx, when combined with unevenly regulated capital, has been poorly constructed buildings that crack, sink, and collapse. Another is less obvious or easy to anticipate – the emergence of a market that exploits this unevenness to take advantage of newly available incentives to rebuild. Developers, sometimes teamed up with individual members of building societies, cash in on increased height allowances, regardless of the impact on residents.

Massive financial incentives to redevelop and a simple online application system for demolition permits have allowed developers to cash in on new height allowances. The globalization of real estate and innovations in the means of securitizing real estate markets have, in the case of Mumbai, upended basic assumptions about the stability of land values and property rights. The phenomenon of Mumbai's falling buildings bears this out quite viscerally.

DIGITAL INFRASTRUCTURES, DISPOSSESSION, AND DISRUPTION

These transformations in Mumbai's metropolitan region raise pressing questions about the financialization of urban space. We argue that the crisis of unsafe and illegal buildings can't be taken at face value. Rather, the crisis is actually a flashpoint in a complex set of transformations in Indian (and global) real estate markets by which cities are being governed through a violent experiment in speculation. Incredible pressure on land values in Mumbai, combined with increased allowances on the height and density of development, has created a fierce economy of redevelopment. Investors stand to gain tremendously from redevelopment, regardless of the actual state of existing structures or the legal and financial implications for residents, leading to extraordinary and tragic forms of displacement and dispossession.

But this is just one part of the story we tell. This story of dispossession is also one of digitization. Financialization and digitization cannot be separated. Indeed, financialization is often explained primarily through histories of state and corporate regulation rather than through the histories (and geographies) of technology. Yet, as Saskia Sassen argues in *Territory, Authority, and Rights,* "the digitization of financial markets and instruments played a crucial role in raising the orders of magnitude, the extent of cross-border integration, and hence the raw power of the global capital markets."[36] The technological capacity for instantaneous exchange is easily taken for granted, but it is a key to understanding not only massive transformations in globalized real estate markets but also events taking place on a single lot in the suburbs of Mumbai.

That we understand this crisis of unsafe and illegal buildings as a question of the digital may be surprising for those who think that digital stories must focus on technologies rather than on their complex embeddedness in everyday life. In a sense, it is easier to think of residents using social media to organize against building demolition as a digital matter. By contrast, it is harder to see the digital in the rise of economic logics, practices, and policies that make these demolitions and the dispossessions they can bring as profitable.

Digital citizenship, as described here, is not simply about how much time one spends on social media or signing petitions online. It is fundamentally about engaging in the vast and often violent transformations to urban life that the digital fuels. Indeed, although the internet monopolizes much of our imagination of digital life, it "is only one portion of the vast world of digital space," and it can be contrasted with the vast, "private dedicated digital networks, such as those used in wholesale

Substandard electrical supply in a Mira Road building. Photograph by Anushree Fadnavis. Courtesy of the National Film Board of Canada.

global finance."[37] Sassen distinguishes these forms from the public internet, explaining that "private digital networks enable forms of power other than the distributed power associated with public digital networks."[38] She also highlights that the leading software development efforts of the last twenty years have been in "firewalled intranets for firms, firewalled tunnels for firm to firm transactions, identity verification, trademark protection, and billing" and not developments that support the decentralizing and democratizing possibilities of digital life that have been widely celebrated.[39]

The financialization of real estate is far from a minor player in the profound entanglement of financialization and digitization. In fact, Sassen uses real estate as an example of the digital and nondigital in the context of financialization to explain the imbrication of these domains more generally. She explores how digitization makes real estate something that can circulate instantaneously through global markets, which financialization today relies on. But digitization clearly does not mean that real estate becomes exclusively virtual: "Financial forms have invented instruments that liquefy real estate, thereby facilitating investment in real estate and its 'circulation' in global markets." Sassen helpfully adds that the nondigital form of real estate persists but is nevertheless affected by its circulation in digital networks. The physical structures that are financialized no doubt remain "part of what constitutes real estate," yet real estate "has been transformed by the fact that it is represented by highly liquid instruments that can circulate in global markets." Real estate "may look the same," she continues, "it may involve the same bricks and mortar, it may be new or old, but it is a transformed entity."[40] Indeed, we could say that one of the implications of digitization and financialization for the nondigital dimensions of real estate in Mumbai is the crisis of falling buildings.

Yet, as we have already suggested, and despite the odds, residents have also made use of digital technologies in their struggles against these events. While this organizing may be new and relatively small-scale, it is nevertheless surprising to us that it is taking shape at all. Although they have middle-class status by virtue of home ownership, apartment owners are a disparate group and often deeply precarious – a small step out of informal housing in the city's slums. In fact, a number of the residents we worked with had moved to their 260-square-foot apartments in Mira Road from informal settlements further south precisely because of their aspirations of upward mobility.

On the one hand, privatized, suburban high-rise living does not produce forms of collective identification or organizing on par with those that have emerged in informal settlements in response to displacement. This practical, material organization of everyday life is compounded by middle-class aspirations and ideologies and by neoliberal responsibilization of the self and the household unit. For suburban high-rise dwellers, home ownership comes with the price of long commutes and long work days, which decrease their ability get to know and work with their neighbours. Finally – the crisis of falling buildings has added crippling debt alongside homelessness to the lives of many of the residents we encountered, further mitigating the potential for collective organizing and action.

While we do not want to overemphasize the power of digital activism, it is notable that it is happening at all. Online activism led to us meeting members of these communities. Although digitization can be extraordinarily destructive, as the crisis of unsafe and illegal buildings shows, the digital can also make things possible, "as even small, resource-poor organizations and individuals can become participants in electronic networks ... It signals the possibility of a sharp growth in cross-border politics by actors other than states."[41]

NOTES

1. Swapna Bannerjee-Guha, "Neoliberalising the 'Urban': New Geographies of Power and Injustice in Indian Cities," *Economic and Political Weekly* 44, 22 (2009): 95–107; and Hussain Indorewala, "Theme Park Mumbai: Hussain Indorewala," *Kafila* (blog), July 12, 2013, https://kafila.online/2013/06/12/theme-park-mumbai-hussain-indorewala.

2. Ashok Bardhan and Cynthia A. Kroll, "Globalization and the Real Estate Industry: Issues, Implications, Opportunities" (paper presented at Sloan Industry Studies Annual Conference, Cambridge, Massachusetts, April 2007), 1, http://web.mit.edu/sis07/www/kroll.pdf.

3. Saskia Sassen, "Mortgage Capital and Its Particularities: A New Frontier for Global Finance," *Journal of International Affairs* 62, 1 (2008): 208–9.

4. Sukhada Tatke, "More Than 55,000 Buildings in Mumbai Illegal," *The Hindu,* June 27, 2014.

5. On expulsions, see Saskia Sassen, *Expulsions: Brutality and Complexity in the Global Economy* (Boston: Harvard University Press, 2014).

6. Naresh Fernandes, *City Adrift: A Short Biography of Bombay* (New Delhi: Aleph Book Company, 2013), 125.

7. Occupancy certificates confirm that a unit conforms to building-code requirements for services and amenities such as power and sewage and to safety standards in design and construction. Though ostensibly required for buyers to take ownership of their flats, in practice the requirement for certificates is rarely enforced. According to an article in *The Hindu,* the number of buildings without them in Mumbai increased by 200 percent between 2010 and 2014, from two thousand buildings to six thousand. See Tatke, "More Than 55,000."

8. Hilary Osborne, "Mumbai World One Tower Flats Go on Sale in London – For a Price," *The Guardian,* February 16, 2015.

9. The legislation (enacted in 1976, repealed in Maharashtra in 2007) was passed to deconcentrate land ownership in the city and, purportedly, enable the state to redistribute land to house poor urban communities. It limited how much urban land could be owned by a single entity.

10. Indigenous communities' status, traditional livelihoods, and territorial rights are recognized in India's constitution.

11 Liza Weinstein, "Mumbai's Development Mafias: Globalization, Organized Crime and Land Development," *International Journal of Urban and Regional Research* 32, 1 (2008): 22–39.

12 One of our informants estimated that 80 percent of Mira Road's buildings had been constructed illegally.

13 According to a *Times of India* article, "Sources said almost all projects executed by Ravi Developers have exceeded permissible limits. In 2009, the builder was pulled up by the high court, which ordered the demolition of Gaurav Gagan in Kandivali (W). The developer had permission for seven floors but had constructed 17 extra floors": Sandhya Nair, "7-Storey Building Develops Cracks, Starts Tilting," *Times of India,* April 30, 2013, http://timesofindia.indiatimes.com/city/mumbai/7-storey-building-develops-cracks-starts-tilting/articleshow/19641565.cms.

14 99Acres, "Gaurav Enclave: Mira Road, Mira Road and Beyond," property website, http://www.99acres.com/ravi-gaurav-enclave-mira-road-east-mira-road-and-beyond-npxid-r20012. One commuter's story is featured in the NFB's documentary *Highrise: Universe Within.*

15 The dangerous-buildings crisis coincided with a wave of activism in which NGOs and citizens used India's 2005 Right to Information (RTI) legislation to scrutinize heretofore undocumented planning, zoning, and building activities. Cities produced counts of illegal and dangerous buildings post-Mumbra in compliance with RTI requests. Some herald RTI as a progressive triumph, while others critique RTI activism as essentially middle-class and neoliberal – focused on state accountability and transparency rather than on equity, redistribution, and social justice.

16 The campaign's Facebook page can be viewed at https://www.facebook.com/SaveCampaCola/.

17 *Highrise: Universe Within* includes a mini-documentary on Campa Cola.

18 It is not clear whether the rendering is of the planned Gaurav Enclave redevelopment, https://www.facebook.com/GauravEnclave/.

19 Her story is featured in *Highrise: Universe Within.*

20 At Campa Cola, our research and documentary team interviewed resident activists, photographed the November 2013 standoff, and worked with the "kids" – as the young campaigners were known – to conduct a survey on how their neighbours were using digital technologies. At Chandresh Terrace, we shared food and stories in several homes and held a community meeting in the compound's courtyard. At the time of our research, the flat owners of Gaurav Enclave had been dispersed for almost a year; we met with two who shed light on the complex legal, financial, and material machinations underpinning the collapse of their homes. We also aimed to connect the three communities; we brokered a strategy-sharing meeting between the Campa Cola "kids" and a young activist in Chandresh Terrace, and we hosted a community forum called "Resident Perspectives on Illegal

Construction, Unsafe Buildings and Redevelopment," at which resident leaders from all three sites came together with scholars, activists, journalists, and planners to discuss these emerging trends and their implications.

21 This is, in part, connected to India's Rent (Control) Act, which freezes rents at unsustainable levels. Even housing-rights activists now point out that this law has, in the end, led to a decades-old standstill of rental-housing development, the complete deterioration of existing rental stock (particularly in the few remaining districts of concentrated rental in inner-city Mumbai), and the demolition and conversion of affordable inner-city rental stock into condominiums and commercial developments. In other words, a law originally intended to provide ownership-like rights of permanency and stability to tenants has instead cleared the way for the rapid redevelopment of inner-city Mumbai via the familiar steps of disinvestment, speculation, demolition, and redevelopment.

22 The predatory speculators, meanwhile, waited to cash in on the newly built flats or to sell their rights as property values fluctuated or the vagaries of the legal processes turned for or against the possibility of a new building ever being built. In Mira Road's real estate boom, they were betting that the long game would pay off.

23 "Illegal Buildings Mushrooming in Mira-Bhayander, Alleges NGO," *Times of India*, September 21, 2015.

24 Of course, collective protections have proven woefully inadequate for maintaining these groups' rights to their lands and homes.

25 Justice H. Suresh, Sudhakar Suradkar, Amita Bhide, Chandrashekhar Deshpande, and Simpreet Singh, "Interim Report of People's Commission on Irregularities and Illegalities in SRA Projects," unpublished report, 2013, copy on file with authors.

26 Tatke, "More Than 55,000 Buildings."

27 Bannerjee-Guha, "Neoliberalising the 'Urban,'" 101.

28 BDD is the Bombay Development Department. Chawls are low-rise multi-unit apartment buildings that house working-class and low-income families.

29 Darryl D'Monte, *Ripping the Fabric: The Decline of Mumbai and Its Mills* (New Delhi: Oxford University Press, 2002).

30 Bannerjee-Guha, "Neoliberalising the 'Urban,'" 103.

31 Saskia Sassen, "Mortgage Capital and Its Particularities: A New Frontier for Global Finance," *Journal of International Affairs* 62, 1 (2008): 187–212.

32 D. Asher Ghertner, "India's Urban Revolution: Geographies of Displacement beyond Gentrification," *Environment and Planning A* 46 (2014): 1558.

33 Ibid., 1559.

34 Ibid.

35 Ibid., 1560.

36 Saskia Sassen, *Territory, Authority, Rights: From Medieval to Global Assemblages* (New Jersey: Princeton University Press, 2007), 336.

37 Ibid., 333.
38 Ibid., 336.
39 Ibid., 331.
40 Ibid., 343.
41 Ibid., 341.

On "Market-Friendly" Planning in Mumbai

Hussain Indorewala and Shweta Wagh

A WELL-KNOWN HISTORIAN of Mumbai once remarked that "at every stage in its history land appeared to be scarce in Bombay."[1] The city's planners have, therefore, looked to the sky to solve its intractable problems – and found an answer in the ratio of built-up area to land area, also known as the floor space index, or FSI. What began in the 1960s as one of the many instruments to set limits on development *intensity*, today gives expression to the city's most promising resource: its air space. FSI has evolved from earlier use as a command-and-control regulation to a *development right* that actively "enables" urban growth; in the hands of urban planners, it is the only legitimate tool to tilt the rapacious arc of the market towards sociospatial "inclusion."

The two versions of Mumbai's latest Development Plan – the Earlier Draft Development Plan (EDDP), released in 2015, and the Revised Draft Development Plan (RDDP), released in 2016 – adopt two different approaches to the FSI. The approach of the EDDP, which was trashed soon after it was released, was to use the FSI as a "tool to guide development" and to cut back on its use as a fiscal mechanism as much as possible. It proposed the expansion of the FSI envelope generally, with variable caps based on the infrastructure availability of different areas (through FSI zones). The planners hoped that by removing distortionary constraints on the real estate market, this "liberalized" FSI regime would create affordable residential and commercial spaces, which would eventually benefit even lower-income groups.

In contrast, the now sanctioned RDDP has reinstated in form and substance the incrementally loopholed regulations of the 1991 Development Plan, which has evolved over twenty-five years through periodic exemptions and modifications. FSI is here conceived as a fiscal tool,

145

allowing higher intensity of development in certain areas by offering it as an "incentive" for redevelopment or as a tradeable commodity – with the condition that property developers build and hand over a part of the new construction for public uses. Although FSI in this case is variable, the differentiation is based on areas specifically targeted for redevelopment, such as slum settlements, rent-controlled areas, and older public-housing units.

Despite differences, both of these approaches have significant similarities. Both make social development contingent on real estate speculation, the value of land being the driver of urban development for the creation of housing, employment, and physical and social infrastructure. Both fail to evaluate the social consequences of FSI increases; higher FSI is offered to higher-density areas either through the allocation of "FSI zones" along transit corridors and inner-city areas (EDDP) or through "targeted" incentive FSI for certain areas and conditions (RDDP). Both aim to promote urban renewal – the first plan through an abstract economic justification, the second through a now artful process of responding to political and electoral pressures by offering concessions to property developers.

Both are also variants of the standard neoliberal dictum: all state protections and regulatory measures distort markets. The cure: the archaic tools of planning must go and, with them, their welfare orientation. As a well-known proponent of this "liberal/pragmatic approach" to planning has explained, planning must be made to work with markets instead of against them.[2] Trying to "figure out the direction or objective of urban growth is a folly"; instead, planners must lay out a grid of streets and, without asserting any hierarchy, accept all sorts of investments. Once they arrive, "You go and reinforce the places and see what the market needs."[3] In this doctrine, land reservations for public use are diminished or abandoned, land-use zoning is liberalized, and building codes are simplified. The state invests in augmenting physical infrastructure, and FSI restrictions are relaxed to stimulate private investment. It also follows that to make these developments viable for property developers, density norms, open space, and environmental standards must be relaxed in these upzoned areas. In short, private enterprise ought to be facilitated at public expense. This approach is based on unflinching faith that, with the right combination of incentives and gifts, private hands will fulfill public ends.

Since 1991, the legal articulation of this approach has been the notorious Regulation 33 of Mumbai's Development Plan. Perhaps the most remarkable embodiment of regulatory capture, the regulation offers additional FSI as an "incentive" to developers and allows "relaxations" to density, setbacks, safety, open space, light and ventilation norms. The general aim of the regulation is to promote redevelopment or urban renewal, and its main casualties have been public-housing colonies, rent-controlled buildings, urban villages, and slum settlements – all the *existing* affordable housing in the city. The thinking goes like this: to enable the private capture of high-value land in urban centres, these built remnants of a redistributionist era ought to be bulldozed, densified, and "rehabilitated," justified by the liberalization-era maxims of "highest and best use" and "inclusive growth."

Meanwhile, the impact of urban infrastructure projects on the landscape of housing in Mumbai has also been striking. Mumbai's Regional Planning Authority (MMRDA) and the Slum Rehabilitation Authority (SRA) have shown how super-high-density, high-rise, and substandard housing for displaced slum dwellers can be made profitable for developers and affordable for the state. Landowners and developers are offered transferable development rights (TDR) to build such housing in low-value areas in lieu of the right to build more in high-value areas as an incentive. According to one estimate, 311,000 households face displacement because of major infrastructure projects in Mumbai.[4] To rehabilitate the displaced, the MMRDA has, since 2001, facilitated the construction of 564 buildings with 64,568 tenements in its various resettlement colonies.[5] Between 1997 and 2018, the SRA has managed to produce 78,901 tenements under its Project Affected Persons (PAP) scheme, and 36,160 are currently under construction.[6]

Asher Ghertner argues that slums "function as a central vehicle for facilitating the alienation of public land to private developers."[7] Although true for high-land-value areas, in depressed areas, resettlement colonies have become active sites for the *generation* of development rights (TDR) for private developers (private land owners surrender their land, and developers construct tenement blocks in lieu of TDR). This TDR is eventually moved and monetized in upper-income residential suburbs, while the slum dwellers who are coercively relocated to resettlement colonies are often those displaced from high-land-value areas. This dual movement provides "lucrative opportunities for accumulation," and its outcome is an even higher concentration and densification of the city's poor.[8]

Curiously, the conception of the slum itself has transformed. Post Independence, formal definitions of the slum approximately coincided with the physical condition of settlements. Conceived as a settlement that deviates from sanitary, building, or land-use codes, the slum was, therefore, a construct of "restrictive" planning norms and practice. Neo-liberal urban policy, which conceives urban land as an exclusive object of private accumulation and exchange and planning as a creative way to elevate property values,[9] has recast the slum as the illegal occupation and suboptimal use of urban land. The slum, in other words, is now a construct of property.

This means that the object of slum policy – and urban renewal generally – has shifted away from environmental improvement towards enforcement of property rights, incorporation of land into formal property markets, and extraction of a share of the development (for the state). The incentive structures for urban renewal, rather than being oriented towards welfare maximization, tilt towards maximizing development *intensity* (to be partially extracted by the state) and *surplus* (to be extracted by property developers). The coalesced interests of state and capital therefore seek to drive down "costs" on development, which include public goods that secure health, safety, comfort, and sociality – making these articles of private consumption. It is therefore unsurprising that high-intensity, high-density construction inevitably produces compromised living conditions for the rehabilitated poor and exclusive private obelisks for the wealthy.

KEYWORDS: THE CHANGING LANGUAGE OF PLANNING

The proponents of the "liberal/pragmatic approach" to planning perhaps draw inspiration from the vice-chairman of the Shanghai Foreign Investment Commission, who argued that urban policy and city planning must aim at "building nests to attract birds."[10] The British economist Nigel Harris has explained that cities must not be planned but *managed,* much like "directing a sailing dinghy." Since a captain cannot control the tides, the winds, or the weather, the best he can do is keep the "consensus of the crew" by continually repositioning the boat to "exploit opportunities" and reach the destination.[11] In contrast to the "old way of doing planning," which stifled growth, strangled markets, and "failed miserably," neoliberal planning discourse emphasizes flexibility, pragmatism, and collaboration. The metaphors and terms of discourse – the worldview within which its adherents "feel at home" – help a community reinforce their social commitments, structure expectations, guide social practices, and rationalize their actions; the changing language of planning can therefore provide a useful entry point into the shifts and continuities of the theory and practice of city making.

Consider a few of the keywords of neoliberal planning.

Vision

A vision is a desired future state but a noncommittal one. Almost every planning document produced in recent decades begins with an ambitious vision for the city: the 2001 McKinsey Report desired "a world-class city," the 2009 Concept Plan desired "elevating MMR to be a Global City," and the RDDP envisaged a "competitive, inclusive and sustainable city." In contrast to post-Independence plans, in which explicit enunciation of planning objectives and the means to attain them are a common feature (despite often falling to do so), vision statements evoke an appealing sense of common purpose, and planners feel no obligation to explain them. Eschewing means-ends reasoning or concern with social outcomes beyond glowing generalities, neoliberal planners champion market rationality and the self-evident propensity of markets to efficiently allocate goods and services and maximize growth.

Frameworks

Frameworks help establish the field within which markets can operate. Faced with uncertainties of the market and unanticipated technological change, any pretense of prior knowledge of or control over the future is "misguided" or "illiberal." The EDDP highlighted the "distortionary" effects of "restrictive" planning as the cause of Mumbai's inability to produce affordable shelter for the poor in Mumbai. It thus proposed to move away from "prescriptive" master planning to a "broad framework" to "allow the market to operate in a competitive manner." The assumption here is that autonomous markets will deliver what is desired without conscious effort; therefore, the role of planning is merely to set broad limits and make fine adjustments where necessary. Attempts to "meddle" in its functioning are distortionary and counterproductive.

Growth

Pre-liberalization-era planning systems, with their public sector orientation, focused on four things: guiding public investment in unprofitable sectors (physical infrastructure); restraining markets where their effects are harmful or destructive (environmental regulations); investing in areas of public interest where market outcomes are seen to be inequitable, unacceptable, or unpredictable (public services, housing, health, education, and transport); and making provisions for those who are excluded from the market (homeless). The planning process was committed, in theory, to redistribution for social stability and equity. Liberalization-era planning processes have shifted the focus to tools that foster economic growth in the hope that the benefits of growth for market actors will trickle down to benefit everyone: the earlier emphasis on *equity* has been replaced by a search for *inclusion*. Everyone has a role and stake in a city, which is constructed as a wealth-producing machine, insofar as the machine itself is not obstructed.

Simplification

"Simplification" is a euphemism for deregulation. Much of the regulatory apparatus of land development in Mumbai evolved at different times around different concerns: fear of epidemic outbreaks, tenant protection, heritage and environmental conservation, land redistribution, and so on. Although plans and regulations were often ineffective in achieving their objectives, their existence signified the state's commitment to safeguarding the interests of the working poor and the environment. Neoliberal advocacy indiscriminately diagnoses these protections as "constraints" and "obstacles" to growth and the cause of the very problems they are meant to overcome.

Partnership and Participation

While "partnership" is used to signal a working relationship with market actors to achieve development goals – implying an automatic convergence between state and corporate interests – participation is often used as a proxy for cooption. A preoccupation with the norms and institutions of participation has occurred simultaneously with the depoliticization and corporatization of governance, the marketization of urban services, and the exacerbation of social and spatial exclusions. Far from any deliberative ideal, "participation" stands for a process that removes any potential obstacles to project implementation: "getting things done" rather than "getting things right."

· · · · ·

Having reengineered cities as machines for wealth creation, neoliberal urbanism has also reduced them to heaps of adversity. Cities have always been sites of social progress, but they are also implicated in the two most devastating market failures of our time: climate change and social inequality.[12] Mumbai is one of a number of coastal cities that will be untenable as the oceans rise; its stark cleavages will inevitably separate those who have many choices from those who have little chance. In the coming decades, the business-as-usual mentality of this "winner-take-all" urbanism will be catastrophic, but it will be a catastrophe that the winners are likely to be the first to profit from, and the first to escape.[13]

The search for spatial justice and the revival of progressive traditions of planning face unprecedented challenges. If planning is the attempt "to link scientific and technical knowledge to processes of social transformation," it will have a decisive role in helping us transition towards urban societies that can secure the well-being of everyone – in the little time we have left.[14]

NOTES

1   Rajnayaran Chandavarkar, *History, Culture and the Indian City* (Cambridge: Cambridge University Press, 2009), 45 (emphasis added).
2   Aparna Piramal Raje, "Bimal Patel: How to Make Urban Planning Work," *livemint*, December 1, 2015, https://www.livemint.com/Politics/NBu03Yn ZHcRSC8r47M1VPN/Bimal-Patel--How-to-make-urban-planning-work. html.
3   Ibid.
4   Simpreet Singh, in "Development, Dispossession and Accumulation: Mumbai in Contemporary Times," *Mumbai Reader 2013* (Mumbai: UDRI 2013), 374–87.

5 Data acquired from the MMRDA through an RTI query in March 2018.

6 Data acquired from the SRA through an RTI query in 2018.

7 Asher Ghertner, "India's Urban Revolution: Geographies of Displacement beyond Gentrification," *Environment and Planning A* 46 (2014): 1554–71.

8 Hussain Indorewala, *Resettling the City: Discriminatory Planning and Environmental Deregulation in Mumbai,* BINUCOM Case Study, 2018, https://mdl.donau-uni.ac.at/binucom/pluginfile.php/405/mod_page/content/32/KRVIA_5_V1.pdf.

9 Samuel Stein, *Capital City: Gentrification and the Real Estate State* (London: Verso, 2019).

10 Peter Hall, *Cities of Tomorrow: An Intellectual History of Urban Planning and Design Since 1880* (Oxford: Blackwell Publishing, 2002), 413.

11 Nigel Harris, "Globalisation and the Management of Indian Cities," *Economic and Political Weekly,* June 21, 2003, http://www.epw.in/special-articles/globalisation-and-management-indian-cities.html.

12 Hussain Indorewala, "The State of Ambivalence: Reflections on the State of Architecture Exhibition at the National Gallery of Modern Art, Mumbai," *Urbanisation* 1, 2 (2016): 195–98.

13 Richard Florida, "Confronting the New Urban Crisis," *Citylab,* April 11, 2017, https://www.citylab.com/equity/2017/04/confronting-the-new-urban-crisis/521031/.

14 John Freidmann, *Planning in the Public Domain: From Knowledge to Action* (Princeton: Princeton University Press, 1987), 38.

Kashaf Siddique on Being Precariously Home in the Suburbs

Deborah Cowen, Kashaf Siddique

OUR TEAM OF RESEARCHERS was fortunate to meet Kashaf Siddique on one of our trips to Mira Road. A suburban municipality to the north of Mumbai's formal borders, Mira Road has been growing at an incredible pace. New construction projects dot the landscape, expanding the boundaries of the built-up area north to Vasai Creek. Urban development in Mira Road has encroached on Indigenous lands and a national park to the east, dispossessing tribal communities and devastating ecologies.

Much of the land undergoing rapid urbanization is made up of tidal flats, places where development plans officially ban construction. Physical growth is not simply a feature of the local economy, it is its fuel. The local economy is anchored in the finance and construction (and sometimes the destruction and reconstruction) of housing. Imported from Bangladesh, migrant construction workers are poorly paid and violently managed. Mafias control the extraction and distribution of the sand and water, which is combined in the concrete that is visible everywhere. Mira's Road's mayor is the daughter of the municipality's largest developer in this real estate fiefdom. Real estate speculation is perhaps the ultimate geopolitical economic game on the periphery.

In some ways, Mira Road is reminiscent of a style of development that has defined the suburbs since they first got their name. Large-scale landowners transform the peripheries of cities into bedroom communities for aspiring homeowners and for large profit margins. The Mira Road rail station brings thousands and thousands from jobs and schools in the south to consume new settlements in the area. The rail line serves as the supply line of demand for use and exchange values. No doubt, the qualities of this particular kind of peripheral development challenge common imaginaries of suburban development. The visual landscape is particularly striking. It is vertical and concrete. Although the logics of urban expansion seem familiar, the rise of finance capital in this context has combined with growing pressures on the land to produce a landscape of extraordinary volume. Density and the suburbs have not been paired in popular imaginaries, but they are inextricably linked in the everyday lives of those who live in the north of Mumbai.

Scholars have devoted little attention to places such as Mira Road; however, events such as those we describe here are being increasingly told in popular and social media. The crisis of illegal buildings that has been widely reported in places such as Campa Cola is in fact far more pervasive on the fringes. Despite the radically different physical form that precariousness assumes in the downtown slums and these suburban towers, there are reasons to consider them together. There may be more similarities than differences in the lives of slum dwellers in the urban core and the people that inhabit this town. Migrants to Mira Road are sometimes the same people displaced by slum clearance initiatives, people who take on extraordinary debt to pursue a purchase. The move to ownership is often supported by the slum clearance programs that move informal communities from the city's core. And ownership, in this context, is a claim, not a fact, since almost 90 percent of buildings in Mira Road are technically illegal. This is not just a semantic point. Residents often bear the brunt of this precarious propertization. That property owners are increasingly facing the violence of demolition once reserved for informal settlements suggests we consider the case carefully.

When our team arrived in Mira Road and at the doors of her building, Kashaf was twenty years old and living with her mother, father, grandmother, and two siblings in a modest apartment building just off a main street in town. Her family was living in a unit that was on the larger side for the complex because it had more than one bedroom. The family had moved to this mixed Hindu-Muslim building in 2010 from Naya Nagar – a stigmatized neighbourhood of Muslim residents in Mira Road that formed following riots to the south.

In 2014, when we conducted this interview, Kashaf was twenty-one and working in ICT. She aimed to complete a business degree, and her younger brother had just started junior college. Their father was a truck driver, their mother a beautician. The move was a leap towards upward mobility, which was reflected in the siblings' career aspirations. Yet these

Chandresh Terrace, 2014. Courtesy of the National Film Board of Canada.

aspirations were threatened by the crisis of illegal buildings and the accumulation of debt that often accompanies them.

Kashaf had become centrally involved in efforts to fight for the rights of her family and the other residents of the complex called Chandresh Terrace. The accompanying image reveals the state of one wing of the building when we visited in 2014. A spontaneous blockade by Kashaf and her neighbours prevented further destruction. Some residents were at home, in the shower or preparing meals, when a demolition company began tearing down the structure. Authorized by the local state, the demolition has no explanation in terms of structural or social problems. The consent of the residents was never established and, in fact, they remained wholly unaware of the plan to tear down their homes until the bulldozers arrived.

The story of Chandresh Terrace, and the efforts of Kashaf and her neighbours, may never have travelled beyond their block or municipal boundaries without the digital revolution. Not only do social media circulate the concerns and claims of precarious populations, of people whose voices may otherwise be difficult to hear – digital technologies also allow residents to document gross abuses by developers and local government and to seek redress, as we explore here. Digital technologies support unusual connections across class and caste; mobile phones and apps such as WhatsApp keep people who are facing parallel crises connected across vast gulfs of social and spatial polarization. Communications technologies have radically transformed research and representational practices.

We met the residents of Chandresh Terrace after following their early organizing efforts online, and we remain connected through Facebook. The exchange that follows was conducted entirely through this medium.

DC: When did your family decide to live at Chandresh Tower?

KS: As we were staying at a rental apartment, we planned to buy a self-owned house by November 2009. After selling a shop we owned, which was in partnership with one of our relatives, we bought this house in 2010.

DC: When did the problems with the building begin? What happened?

KS: In June 2013, a few members of the building society were secretly planning to redevelop the complex, and thus gave wrong details and set wrong expectations in residents' minds. These members tried to manipulate residents and succeeded in doing so with maximum people of the society, but some residents including us went ahead and investigated. We found that the developers were not professional developers, in fact they had teamed up with the committee members with a goal to cheat the residents.

These building society members were planning the forgery since 2010 when we shifted to the complex and no one else in the building had a clue about what was going on. Later, Annual General meeting papers were circulated in the society and most of the residents signed agreements without knowledge as most of them were illiterate. These papers were used against us. Committee members went ahead and filed a complaint with the municipality saying that the building is unsafe to live in. A fake report saying the same was created.

Suddenly on 21st August, police officers from Kanakia police station along with some female officers came into the society with the contractors and started demolishing the 1st Wing. Power supply, water pipelines were cut down and supply for both were stopped. As maximum of the residents had signed the AGM papers, they accepted the contractors' offer and started moving from Chandresh Terrace, taking their deposit and rent for their rental apartments by trusting the committee members. But somehow ten of the residents were still living [there] and did not move out and were in process of shifting.

Meanwhile, A Wing was completely demolished except for three apartments because they were still residing and were against the committee members.

The rest of the wings were not demolished as residents teamed up and opposed them, confronting them as they moved to demolish the other wings. As it was very sudden, a political member walked in trying to help us. He helped us in calling commissioner again and requested to investigate the remaining wings and create a new report about whether the buildings are safe to reside in or not. With this support of commissioner, power and water was restored after a week's time. Residents paid "Reliance" for restoring the same.

After a month a report was sent by the municipality stating that building only needs to undergo some repairs and then it would be totally safe to reside. Then after this, a case was filed against the municipality and the developers. The case is still pending with the court. Builder here is unauthorized and is not ready to back off because they have already invested in the project. Builders did not have any office nearby it was approximately 50 km's away. They were five partners but they did not have any bank security. Some clauses that are required to redevelop a building were also not fulfilled.

Most importantly "government resolution" is required to redevelop any building and this permission was not provided by the government for our complex. Government Resolutions are only provided as of now to buildings built before 1985 whereas "Chandresh Terrace" was build up in 1993 and was registered in 1997 as informed by the committee members. So government resolution was exceptional and impossible in this case.

Seventy percent of residents being illiterate was a main reason for this horrifying situation. There was no digitalization and less socialism as 80 percent of the crowd belonged to different parts of

India. The residents were from South and West but the maximum of them were from the northern part of India. But this incident bought us together as a team of fighters, fighting for their rights, and is still fighting for the rights to be received.

A year later, the builder stopped providing funds for rents to residents who had shifted trusting the committee members. As the residents were now irate and left with no option and after a lot of group meetings among residents brought all of us together. Residents from "A" wing did not return the deposit amount to the builders and utilized the same to rebuild their particular apartments and started reliving there. As the case is still going on with the court and no resolution has been provided yet, power and water supply for these residents could not be restored. Hence the rest three wings B, C, and D are sharing the power and water supply with them.

Yet we haven't received any resolution from the court. The case is still pending with the court and as left with no other option we are waiting for the results. We hope for a better future of our society and hopefully wanting to see some miracle to hit us.

DC: When did residents start getting organized?

KS: October 2014, residents started moving back to Chandresh Terrace building up their apartments, sharing electricity with the other wings and society pump house, arranging water supplies for them using the deposit amount given by builder.

DC: When did you get involved in organizing with other residents? Why?

KS: As they lost support from builder for monthly rental and when they had their own apartments here, why somewhere else? We had a society meeting for this where everyone stepped ahead supporting them to build up again with all of us. As staying united will save our living.

DC: What kind of work have you done? How often do you meet with other residents to talk about the problems you face or to work towards solutions?

KS: We all have been together working together nothing has been done single-handedly. We have been running in and out of courts and next court dates which is still pending. We have been using political contacts to build pressure. We have bought up a letter from Hasan Dalvai to re inspect Chandresh Terrace and then to come up with decision where the reply came out that the building only needs repair rather than taking them down. Funds needs to be raised for them and is still under process. Hoping for the best outcome. Meetings are now conducted real seldom. Yes, but we residents do meet up every alternate week to discuss and resolve if any issues.

We have been connected through WhatsApp and messages all over. I personally have been sharing videos on demolished buildings and pictures for same. Did not get much support though. Internet here is less involved, people here are not much educated and not much connected with advance technologies. This has been a drawback, I feel. Due to less socializing this problem did not reach out much. Else maybe it would have been different by now. Cellphones have played a great role here to capture pictures for ongoing thing such as demolition; meetings conducted were also recorded with the help of cellphones. That helped a lot.

Everyone here is busy making living for their own, but still this has got us lot closer emotionally. We have been supporting each other now for every small thing, sharing things going on with building and society. Now we usually meet, greet each other unlike before.

I see all of us united and together now. And that is something what matters the most. This helps us fighting everything around. These happenings around me has built me up as a strong person and as a fighter, who will never give up until the end.

Dispatch from Mumbai

Deborah Cowen, Paramita Nath

THIS DISPATCH WAS written in late November 2013, during our team's first visit to Mumbai, and updated in May 2014, after the Supreme Court issued its decision. Residents of the compound were forced to move out of their flats in June 2014, after a demolition order was issued, and they returned in 2015, after that order was overturned by the apex court. In 2020, the time of publication, the community remained in place, though the legal status of the compound is still before the courts.

At 11:00 a.m. on November 11, 2013, Mumbai's municipal government prepared for yet another mass eviction of its citizens. Hundreds of police barricaded the streets, followed by hundreds more from the demolition team. Oversize military-style paddy wagons and massive bulldozers rolled into the area, met by residents and a sea of reporters from all of India's major news outlets. Residents were arrested, along with a number of politicians who had joined them in their fight.

Already garnering front-page coverage across India for weeks, what made this incident standout from the long string of violent evictions in this city was the social standing of residents. Campa Cola is not a slum settlement but a largely middle-class apartment complex in a gentrifying neighbourhood of the city.

A surprising target for demolition, the story of the Campa Cola complex is increasingly common. Recently revealed is the astonishing fact that almost half of all construction in Mumbai of the last decade is technically illegal. The exact figure, obtained by activists through India's recent Right to Information Act (RTI), is staggering. More than 56,300 buildings in Bombay hold this same status, and so, in theory, could face a similar fate.

Demolition day at Campa Cola, November 6, 2013. Courtesy of the National Film Board of Canada.

RTI is one impressive new technology in the struggle over urban land in a country where corruption is common and authority often arbitrary. But it is not the only new tool – social media has also become a powerful weapon. The youth of Campa Cola led an elaborate campaign on Facebook, Twitter, and WhatsApp that catapulted their saga to the centre of national public debate. These well-connected residents and their tech-savvy kids mobilized the support of a wide cross-section of Indian society. Through their Facebook page, people from as far away as Dubai, Iran, Australia, the United States, and the United Kingdom also signed the petition and supported the campaign. Official support from all the major political parties followed.

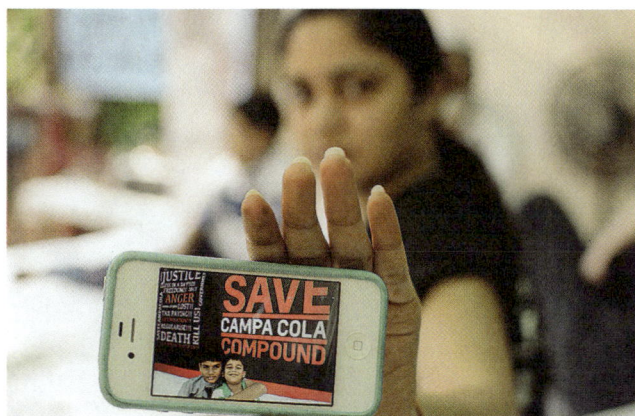

The social-media campaign designed by the Campa Cola "kids" used sophisticated design and marketing. Courtesy of the National Film Board of Canada.

▲ **Campaign headquarters at Campa Cola.** Courtesy of the National Film Board of Canada.

▶ **Condo rooftop.** Courtesy of the National Film Board of Canada.

▼ **The aging buildings of Campa Cola against a backdrop of newer development.** Courtesy of the National Film Board of Canada.

And yet, despite this level of support, a dramatic standoff between residents, on the one hand, and state security forces and demolition crews, on the other, lasted for hours as the dense morning smog lifted from the Worli neighbourhood.

If the figures of Mumbai's illegal buildings are surprising, the story becomes even more staggering in the rapidly spreading suburbs that surround the city. In areas like Mira Road to the north, more than 80 percent of buildings are "illegal," and a whole economy is emerging around the demolition and reconstruction of apartment towers that are often only a few years old. Ironically, these suburbs are increasingly home to people displaced from the city through forced evictions tied to urban-renewal schemes, by skyrocketing real estate prices that are now firmly inserted into globalized land markets, and by, historically, the riots. Displaced from the downtown, residents are often then displaced from their new vertical suburban homes, only to find themselves temporarily re-housed in buildings that are also illegal. High-rise residents have become pawns in an elaborate economy of illegality.

Far from exceptional, Campa Cola's illegal status has become something of a norm in the new Mumbai. It is a product of real estate capital run wild, of staggeringly high rents caused by international financial speculation, of deepening economic and political polarization, and of deep-seated corruption in a city where a powerful "builder-authority nexus" acts as judge, jury, and executioner in the making of urban space.

Bold proposals from municipal and state authorities and their corporate partners, taking funds and directives from the World Bank, aim for total transformation of the city's built form and social geography. With lofty visions of transforming "Mumbai into Shanghai," the city's concept plan proposes massive across-the-board hikes in density to accommodate an imagined population of 44 million by 2052. Increasingly, slum settlements are being transformed into vertical spaces through schemes that put high-rise "slum rehabilitation projects" in their place, regularizing their legal status. In fact, the Campa Cola site abuts one such project, Maya Nagar, in which residents have been moved, in phases, out of an informal settlement into a temporary "transit camp" and then into newly constructed high-rise apartments. Many Maya Nagar residents do domestic work in Campa Cola homes; paradoxically, their employers implored the former slum dwellers to support the Campa Cola campaign.

The trend to regularize informal settlements makes the Campa Cola case even more curious. The official explanation for the Campa Cola towers' illegality is floor space index violation (known simply as "FSI violation") – meaning that they exceed the allowed density for the site. The municipality demarked everything above the fifth floor for destruction. For some of the buildings, this would mean demolition of a handful of floors, while other towers will see as many as twelve and fifteen storeys shaved off the top. All this demolition is under way even as authorities engineer plans for a newly and intensely vertical urban form.

Residents learned the details of their buildings' illegal status in 2005 – more than twenty-five years after their construction. They have been living in the complex since that time, making the suddenness of the situation absurd and the timing questionable.

Why this building? Why now? Until recently, the municipality was happy to collect taxes and look the other way. But residents point out that things changed when a developer purchased an adjacent lot. This neighbouring site, never legally severed from theirs, has no density allowance by virtue of the overdevelopment of Campa Cola. The developer next door stands to gain the density to be recycled through demolition. Inheriting FSI is as good as the gift of gold. The language of "land grab" circulates widely in Mumbai these days.

▲ **Police face off with residents on demolition day.** Courtesy of the National Film Board of Canada.

◀ **The Maya Nagar transit camp adjacent to Campa Cola.** Courtesy of the National Film Board of Canada.

▶ **The standoff.** Courtesy of the National Film Board of Canada.

After elaborate protests and legal challenges, the Supreme Court issued a stay delaying decisions for seven months, leaving the fate of the complex unclear. Renewed eviction notices were delivered to the residents, with a deadline of May 31, 2014, to vacate the premises. The Supreme Court declined a further stay on evictions, but residents have been granted one more chance to make their case before the Supreme Court on June 3.

The ironies of illegality are bitter. Despite lip service and a proliferation of policy, little is done to shelter Mumbaikers from a string of recent building collapses – most of which have been "illegal" buildings or those built to house the many millions who live on streets and in slums.

▲ **Campa Cola residents try to save their homes.** Courtesy of the National Film Board of Canada.

▶ **State forces assemble.** Courtesy of the National Film Board of Canada.

Without a doubt, the violence committed against people living in poverty in the struggle over urban space is brutal – actual rather than potential – and often lethal. Slum settlements of many thousands have been razed repeatedly without any significant media attention or parallel consideration by the courts – but so, too, have suburban towers that house a growing share of the city's population, including large numbers of the new suburban poor. There are no guarantees that the intervention that could come on behalf of the Campa Cola residents would in any way benefit poor people in similar situations without valiant political organizing. Nevertheless, the Campa Cola conflict and the epidemic of illegality are important precedents in this megalopolis, where the vulgar use of law and land for profit is sculpting the future shape of the city.

#WhyLoiter

Shilpa Phadke, Sameera Khan

THE TWENTY-FIRST CENTURY has seen a growing focus on crimes against women in public throughout the world. India has shown a heightened interest in ensuring women's safety in public spaces, particularly since a young woman was gang raped and murdered on the streets of New Delhi on December 16, 2012. The incident met with angry protests across the country and expanded media coverage of cases of violence against women and rape in particular. Never before in India have the issues of women's safety and gender violence been so mainstream. They are discussed by people who don't identify as "feminist," and they were an agenda item in the manifestos of political parties in the 2014 national elections.

Other rape cases have engendered intense media coverage, notably that of a young journalist raped in Mumbai in August 2013 (the Shakti Mills case) and that of a woman assaulted by her Uber driver in Delhi in December 2014. In these cases and the media, there has been a tendency towards victim blaming and shaming, and this tendency is neither new nor unique to India, or South Asia, as reportage of sexual assaults across the world so eloquently demonstrates. When the questioning in the Uber cab rape case turned towards asking the victim why she had fallen asleep in the taxi after a night out, we decided it was time to roll out a campaign outlining women's right to fun in the city. The campaign followed and was layered onto the multiple workshops and sessions we had been conducting on women and public space in Mumbai and elsewhere in India.

Our book, *Why Loiter? Women and Risk on Mumbai Streets,* co-authored with Shilpa Ranade, was published in 2011. Our main argument was that loitering was indeed the way forward in making claims to the city. Following the rape cases, we were inspired to continue this work through an online campaign called #WhyLoiter, which ran between December 16, 2014, and January 1, 2015, inviting women to use the hashtag #whyloiter to post status updates, messages, and photographs on social media of themselves or their friends having fun in public spaces in cities.

We wanted to make the point that women were already accessing the public to loiter for pleasure. Our hope was to fracture in some small measure the overarching mainstream discourse of fear and danger when it came to women's presence in public spaces. December 16, 2014, also marked the two-year anniversary of the gang rape and murder in Delhi. The campaign was a response to the fact that discussions of women in the city had almost exclusively become about violence. We wanted to switch the focus of debate to women's right to the city and, just as important, women's right to take pleasure in the city. The idea was to create a sense of a community among women in public spaces so that we can remind ourselves and other women that we are not alone.

We began with four posters created by advertising professional Nishant John and designer Abhishek Jayaprakash.

You don't always have to stand up for justice. Sometimes, *sitting down works.*

Make a statement about women's right to loiter in public space.
#whyloiter

We then wrote out a set of tweets to start us off and asked Neha Singh and Devina Kapoor to participate. They were women who had read our book and, along with a growing tribe of women, had taken on the task of actively loitering in the city every weekend. Soon after we started publicizing #WhyLoiter, the campaign was flooded with posts.

DHRUBO JYOTI @DHRUBO127: #whyloiter outside parliament street police station. Cuz women have the right to be anywhere, do anything. 2:54 PM – 16 Dec 2014

LAWYERETTE @CHHOTI_VAKEEL: I like to stand by the road, drink tapri chai and watch the traffic whiz by. #whyloiter December 16, 2014

LITTLE ROCKET GIRL @LILLROCKETGIRL: Because I wish to loiter. #whyloiter, 3:50 PM – 16 Dec 2014

SHRUTI RAVI @EVEANTIUM: After 8 years of negotiating, this year my folks finally agreed to not have a set curfew timing for me to reach home by. #whyloiter, 11:28 AM – 16 Dec 2014

HIMALA JOSHI @HIMALA: For the peace and pleasure in walking. Early in the morning, or late in the evening. #whyloiter @whyloiter, 10:04 AM – 17 Dec 2014

NEHA DIXIT @NEHADIXITI23: Thanks #whyloiter for reminding me why I need to be out for more often & for no reason. Dec 18, 2014

SHUBHRA GUPTA @SHUBHRAGUPTA: #WhyLoiter? Just because. Because I want to. Because I feel like it. Jan 6, 2015

SWATI VIJAYA, BLOGGER: Why Loiter? as a book and a campaign for me and many other women effectively articulated the importance of the guilt-free right to access public space that we should claim and brazenly practice. More power to women who loiter, linger, dawdle, dally and wait!
  – "Of Waiting, Loitering and Sulking," *Larrikan Letters* (blog), Jan 2, 2016, https://larrikinletters.wordpress.com/2015/01/02/of-waiting-loitering-and-sulking/

CLAIRE MOOKERJEE @CLAIREMOOKERJEE: #WhyLoiter sought to visibilise women's fun in public space. Women should loiter in public space everywhere to change attitudes #whyloiter? Mar 10, 2015

Kunal @kunalone, IAS, District Magistrate East District: Read about #whyloiter campaign. Its nice idea to democratise public spaces and reclaim these for all specially females 24 × 7. April 14, 2015

Responses came not only from the big cities such as Mumbai, Delhi, Kolkata, Hyderabad, and Bengaluru but also from smaller towns and diverse locations, from Sikkim, Patiala, Jammu, Bareilly, and Bhuj, all the way to Bhilai, Mangalore, Kochi, and Manipal. Many women from outside India also got interested in and participated in the campaign. Women, alone and in groups, sent us photographs of themselves hanging out in public spaces and partaking in the simple pleasures of being in public – walking by the sea or a river, sitting on the grass in a park, drinking chai on a street corner, and wandering the city in a cycle rickshaw. Even older women sent photographs and messages, one of them mentioning that having tea in a *tapri* (street stall) was a pleasure she had denied herself all her life because it wasn't something "girls from good families were supposed to do."

As the days went by, we added posters and messages. We created word clouds representing the thoughts of women in particular cities, and we started a crowdsourcing initiative to put together a list of words for loitering in different languages.

#WhyLoiter certainly started a conversation about women, public space, rights, cities, and fun. Some of those ideas travelled, and new campaigns that complemented ours emerged. Soon after our campaign ended, we collaborated with Blank Noise's bold campaign, Meet to Sleep, which invited women to stretch out and take a nap on a park bench or the grass. We organized two sessions in different parks in Mumbai. In 2015, we also witnessed the growth of two exciting South Asian movements to claim the city for citizens. One of them began with a Tumblr photo blog called *Girls at Dhabas*. The movement originated in Karachi and describes itself as "Desi feminists and women defining public space(s) on our own terms and whims." It encourages Pakistani women to hang out at public spaces such as *dhabas* (street-side tea shops usually haunted by men) or to take over streets for an afternoon game of cricket (as men do).

The other movement, the #PinjraTod (Break the Hostel Locks) campaign, was started in Delhi by women university students. As their Facebook page states, the campaign was a collective effort to take on "the discriminatory rules and regulations (at colleges and hostels) that seek to control and limit the access, mobility, sexuality and experiences of women who come to study, work and live in the city." They have

Fe   a   a a h)

Taporigiri (Bambaiya)

Lafua (Bhojpuri)

Gedi (Punjabi)

Flâneur (French)

Faraylaaey-helo (Kutchhi)

KHATTI (ODIA)

Darbedari (Kashmiri)

L   g (glish)

Hang-gai (Cantonese)

Rakhadpatti (Gujarati)

Awaragardi (Hindi)

Othla-Hodiyodu (Kannada)

Thenditharam (Malyalam)

Bariteginchindi (Telegu)

Chorki-kata (Bangla)

Alayarathu (Tamil)

Tafrih (Urdu)

Herumhaengen (German)

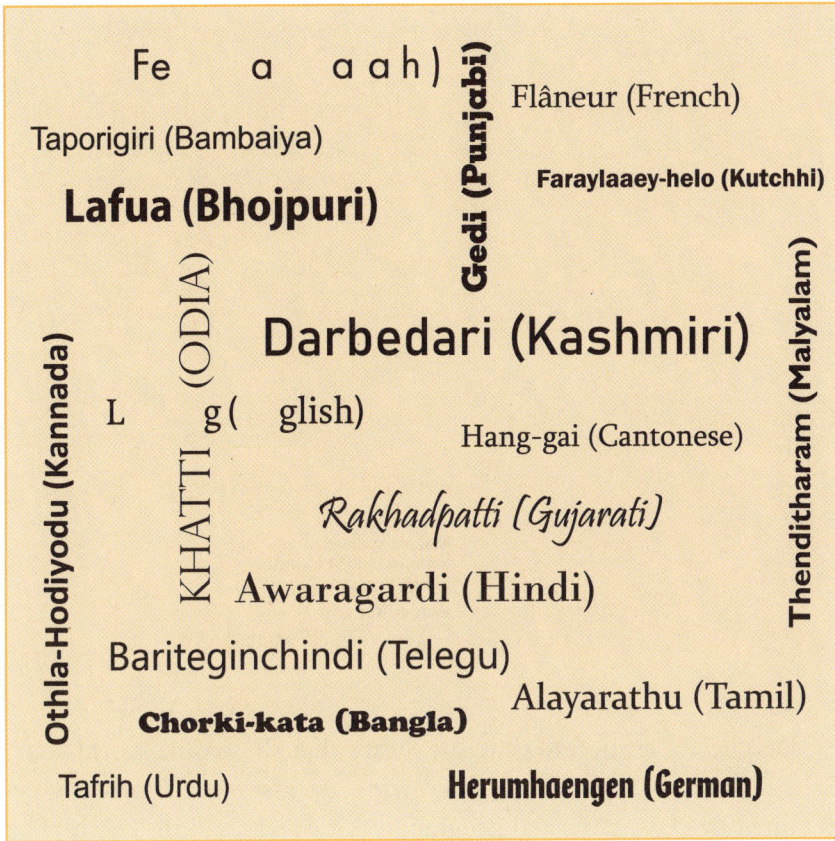

Words for loitering in different languages.

undertaken night walks and all-night study sessions outside of university libraries (from which women hostellers are often barred because of early curfews).

#WhyLoiter's online presence has enabled us to make connections with these groups and others, building bridges across borders. We ran another energetic campaign from December 2015 to January 2016, and this time it involved multiple groups and evoked multiple conversations. One of the most exciting outcomes was that a substantial part of it took place across the border in Pakistan, where Girls at Dhabas initiated loitering in parks in Lahore, Karachi, and Islamabad. The Pakistan chapter of the Fearless Collective, a group based in Bangalore in India, also participated.

Not all
protests
are marches.
Some are
strolls.

Make a statement about women's right to loiter in public space.

`#whyloiter`

Despite these successes, we are aware that the online world, especially in India, is limited though growing. Access to social media is determined by class, caste, and geographical location, and the reach of an online campaign is limited at best. Interestingly, the mainstream media, including the non-English language press, reported widely on the campaign, somewhat broadening its reach. We also understand that online campaigns tend to be less nuanced then other types of campaigning, especially on Twitter, where everything must fit into 140 characters. Notwithstanding these limitations, in its aim to generate a wider debate on women's right to fun in the city, the campaign was largely successful. We really did not have to do much to propel it – women already had the desire and the capacity to access the city and enjoy its delights. We simply provided them with a space to articulate their thoughts and demonstrate their actions.

Shifting and
Scripting
Urban Lives

High-Altitude Protests and Necropolitical Digits

Ju Hui Judy Han

DATES. AUGUST 15, OR 8.15, marks the day of Korea's liberation from Japanese colonial rule, and 6.25 is the name used by South Koreans to refer to the Korean War. 4.3 is the day the Jeju massacres took place, though the killings took more than just one day. 4.19 refers to the student-led uprising in 1960 that toppled the dictatorship of the First Republic, and 12.12 was the day of the military coup d'état in 1979 that created the Fifth Republic. 5.18, the day of the Kwangju uprising in 1980, marks the heart of prodemocracy movements. Korean Americans say 4.29 or *saigu* when referring to the LA uprising in 1992.

These numbers memorialize beginnings and ends. They are temporal abbreviations, digits encoded with meaning, historical moments compacted into acronyms. Calendar dates repeat every year. Anniversaries fasten one year to another, insisting on continued remembrance, grief that does not end. Every year, on 4.16, the Sewol ferry sinks with 304 lives aboard.

Another political cultural practice in Korea accounts for both persistence and impermanence of time and space: numerical measures of endurance. It took 1,895 days for Kiryung Electronics workers (2005–10) to win reinstatement, but they returned to work only to find an empty building; the company had relocated in secret without notifying the workers. Korean Train Express attendants took 4,526 days (2006–18), Jaeneung Educational Institution (JEI) workers took 2,076 days (2007–13), Korea Telecom workers took 517 days (2000–02), and the list goes on. Former "military comfort women," women who were drafted into sexual slavery for the Imperial Japanese Army, have held a protest rally outside the Japanese Embassy in Seoul on every Wednesday since January

8, 1992, for a total of 1,441 weekly protests (as of May 27, 2020). It's been over twenty-eight years and and it is still ongoing.

Digits also mark the duration of high-altitude protests, sometimes known as aerial protests of *hanŭl kamok* or prisons in the sky. In militant labour politics – where long, protracted fights are a mainstay – high-altitude protests simultaneously perform risky near-death precarity and a refusal to succumb. Jinsook Kim spent 309 days atop a construction crane in 2011 in protest against Hanjin Heavy Industries and Construction. She chose Crane No. 85 after fellow activist Kim Ju-ik hanged himself on it on the 130th day of his occupation protest in 2003. In 2012–13, two auto workers lived for 296 days on a high-voltage transmission tower in Ulsan to protest against Hyundai Auto across the street. That same year in Seoul, two women, Yeo Min-hui and Oh Soo-young, protested for 202 days on top of a Catholic church bell tower to demand reinstatement and justice for private educational workers at JEI.

Usually alone but sometimes in pairs, high-altitude protesters appeal not only to the rule of law but the law of gravity and fear of death. They engage in a high-rise spatial fix to dramatize the frailty of life, not courage or strength. They wave and tweet encouraging words to family and supporters on the ground, but up in the sky, in that ghastly isolation, high-altitude protesters manage exhaustion and monotony, anxiety and creeping depression. Death is always only a small step away. They fight the constant urge to fly.

"Kogong'yŏjido" is a map of high-altitude occupation protests designed by Park Eun-sun of the urbanist artist-activist collective Listen to the City. Originally published in tabloid newspaper size, the map is accompanied by a long list of the starting dates and locations of the 116 labour protests that took place between 1990 and 2015. The map shows how tall the structures are, in metres, and the total number of days survived on these monuments to modernity, heaps of metal and concrete. For over twenty well-known cases illustrated on the map, architectural line drawings render the physicality and materiality of high-rise infrastructure that underpin high-altitude protests – construction tower cranes, suspension bridges, factory chimneys, and billboards. Protesters climb up and set up camp, spending days and nights in sweltering heat and monsoon rain, in snow and ice, separated by distance but also connected – hopefully, but not always – to support networks on the ground. For water and food, for waste disposal, and for devices, through

고공여지도

高空輿地圖

1990-2015 전국 고공농성 연대기

"Kogong'yŏjido," a map of high-altitude occupation protests in Korea, 1990–2015.
Park Eun-sun, Listen to the City, www.listentothecity.org.

digital devices, comrades and families on the ground constitute a lifeline. High-altitude protesters may be perched up in the air, but they are kept alive by their connectivity to the ground, illustrating the biopower of life-sustaining solidarity stretched vertically across space.

On July 9, 2015, after spending a world record–setting 408 days on a forty-five-metre-tall factory chimney to demand reinstatement, Cha Gwang-ho, a dismissed Star Chemical/FineTek worker, carefully climbed down to a small cheering crowd. He headed up on May 27, 2014 and came down on July 9, 2015. Immediately after a cursory health check, Cha was taken into police custody and jailed for charges of obstruction and trespassing in what was described as a seamless transition from one prison to another. Cha's record was broken on January 11, 2019, by his coworkers, Hong Gi-tak and Pak Jun-ho, who came down after surviving 426 days atop a factory chimney seventy-five metres high, setting a new world record for high-altitude occupation protest. Weakened from fourteen months of occupation and many days of hunger strike, Hong and Pak took fifty minutes to descend.

Encoded in the digits of high-altitude protests are not triumphant narratives of achievement but profound protest performances of necropolitics – they are battling the constant fear of falling, fear of death, and fear of being forgotten. They demonstrate what might appear to be groundless human resilience against all odds.

Terabytes of Love

Indu Vashist

> Tera, tera, tera, byte-ah kadhal irukku
>> *I have tera, tera, tera, terabytes of love for you.*
>
> Neeyum bit-u bit-u bit-ah byte-u panna yerum kirukku
>> *It drives me crazy when you byte me bit by bit.*
>
> Instagram-athula vaadi vazhalaam
>> *Come, let's live in the Insta-village,[1]*
>
> Namma vazhum nimishathellaam suttu thallalaam
>> *and let's capture every minute of life.*
>
> From "Selfie Pulla,"
> performed by Vijay and Sunitha
> Sarthy and featured on the
> *Kaththi* (2014) soundtrack

LATE IN DECEMBER 2014, I walked out of the Chennai International Airport after being away for two years. It was 5 a.m. and a balmy twenty-seven degrees. The breeze carried the smell of the ocean mixed with the nearby garbage dump and construction. As I got into a government taxi, with the prepaid fare slip in hand, I turned to my partner and said, "Being away from India for two years is like being away from Canada for a decade." The pace with which things change is unimaginable to the Western mind.

Whole areas of the city that hadn't been built before were now bustling. In Velachery, we passed the largest mall in Asia, which wasn't there the year before. On Velachery Main Road, now a highway, people from the new apartment towers walk, jog, and cycle before the workday begins and the sun starts beating down. As we turned into the IT Corridor in Perungudi, a small village in the Chengalpattu District that was absorbed into Chennai, we discovered that the road leading to our house, which had been a potholed dirt road during my last stay, was now a paved, perfectly smooth surface with a bus route, a hip yoga studio, and four new parks and playgrounds.

The congested Chennai streets were littered with cars marked as TaxiForSure or Ola Cabs. We asked the taxi driver to turn on the radio, and the superhit of the winter, "Merasalaayitten," started blaring. In the video for this song, represented as a flashback in the film *I*, the hero holds a Nokia brick phone, which magically comes alive and turns into the heroine. She wears a blue-and-black body suit, the colours of the mass-produced Nokia cellphone. The camera is between her shoulder blades, the green and red "On" and "Off" buttons are on her breasts, and the numbers are on her torso. The hero proceeds to push her buttons.[2] This song played incessantly on the radio for three months.

As we walk into the house, fresh steaming *idlis* and filter coffee are on the table, and my partner's mother is ordering a taxi on her smartphone. Even before she asks about our journey, she starts raving about the savings offered by TaxiForSure. "Can you believe that the driver turned on the metre? I went to Thiruvanmayur in just eight-eight rupees!"

I inserted my prepaid subscriber identity module (SIM) card into my trusty brick phone (with FM radio!) and couldn't pick up a network because the connection had expired. I walked to the local cellphone *kadai* (shop), where I meet Rahman, a young, fluently trilingual migrant worker. He breaks down all of the packages available. There are data packages specifically for the avid WhatsApp user and a social-media package in which Facebook and Twitter usage are free. He takes one look at my brick phone and laughs. He asks me how I expect to run apps on it. He directs me to the nearest cellphone showroom to get myself a smartphone.

In 2014, India has more than 300 million internet users, making it the second-largest internet user base in the world. While this looks like a giant number, it makes up approximately only 20 percent of the population. An industry report projects that there will be 500 million internet users (out of a total population of 1.25 billion) by 2017, the vast majority of them mobile users.[3] This immense growth is attributable to the availability of inexpensive smartphones and low-cost data packages.[4]

The target market of this growth is the hinterland. The Idea network (India's leading telecom provider) released an ad in which a woman in rural Haryana is shown in seclusion. She says, in a local dialect, "In my village, girls are not allowed to go to college. I asked them, 'What if college came to us?'" The scene cuts to a group of women studying in seclusion on their smartphones and speaking in English. The ad ends with her praying to her cellphone.[5]

This expansive growth of internet usage is rapidly changing the fabric and imagination of Indian society. The hit Tamil song "Selfie Pulla" illustrates that technology has crept into common parlance and has been absorbed into existing tropes. The proliferation of smartphones and affordable access to the internet have meant an explosion of apps on the market. There are apps to order food, to pay bills, and to order taxis. The ever-present nature of technology in society has not only bred curiosity and excitement about the internet but also the desire to adapt and work with it.

After I settled into Madras and started using the new app-based taxi services to get around, I became curious about the workers behind these apps and how they understood their work. To unravel the connection between the internet, technology, and work, I decided to go directly to the source, the labourers whose work is now mediated by the internet. At the time, there was a fierce competition between TaxiForSure and Ola Cabs to monopolize the app-based taxi market in Madras. Billboards advertised each app's rates, and newspaper articles marvelled at the savings being passed on to the consumer.

I interviewed auto rickshaw drivers working for Ola, and I scoured the business papers to understand how these businesses could afford to constantly undercut each other and still survive. Ola Cabs is an indigenous Indian personal transportation aggregator app and website that provides users with access to a variety of vehicle options, ranging from luxury cars to auto rickshaws. The auto rickshaws were incorporated into the Ola system three months prior to my interviews. Ola controls over 60 percent of the transportation aggregator market in India.[6] Despite failing to break even, it is considered one of the fastest-growing businesses in the country because it has financial backing from venture capitalists based in Silicon Valley.[7]

During the time that I interviewed the drivers, they were offered new incentives by Ola. Initially, Ola was willing to lose an average of two hundred rupees per trip to compete with TaxiForSure and Uber.[8] Ola managed to raise enough funds to drive TaxiForSure out of business and acquire their workers and technology. The Competition Commission of India has ordered a probe into Ola on that grounds that it "has incentivized drivers unrealistically by using the money it has received through foreign investments, including from Tiger Capital, which can never be matched by the existing radio cab operators or potential indigenous enterprises desirous of starting such operations in India."[9]

I interviewed the drivers, with translation help from my friend Anusha, while performing errands around Chennai. I chose them at random. I logged on to Ola and ordered the nearest auto. I told the drivers I was interested in their experiences working with the Ola app. Each interview lasted from fifteen to thirty minutes. All of the drivers had come to own their vehicles after renting for most of their careers. So they were the direct beneficiaries of all their income. Their experience with the app ranged from a few weeks to three months. According to the drivers, when the app had launched three months earlier, Ola recruiters had visited the roadside auto stands to recruit drivers. After the first batch of drivers signed up, the rest were recruited by word of mouth.

The drivers were given a device that looks like a smartphone to run the Ola app. This device contains a SIM card, which allows it access to the app, which in turn connects the driver to their clients. The device also contains a GPS system and the Ola wallet app. The drivers are also given a bank account with ICICI bank, which is then synced to the Ola wallet. The device cannot perform any other functions, including phone calls. When the drivers reach the destination, or as close to the destination as the GPS will allow, they must call the client to indicate that they've arrived or ask for directions. The costs of these phone calls are borne by the driver.

Across the board, drivers and clients in Chennai prefer to use Ola over hailing roadside autos because of the ease of the transaction. While the state government had mandated that metres be installed in cabs, the base rate has not increased to reflect the rising cost of living.[10] The practice is to haggle with drivers until a price is agreed upon. By contrast, when booking an auto with Ola, the client knows that the driver will turn on the meter and that they must pay ten rupees above the metered rate. The informal system of bargaining is an everyday stress for both drivers and clients. One interviewee commented, "I don't argue with anyone. What I like about Ola is that there is no argument. I don't have to ask for plus twenty or plus thirty [above the meter]. This is convenient. They just give the rate that shows up in the app. One lady was telling me earlier, normally, when you go to hail an auto, you hail three or four, and then only the fifth one will be willing to both take you at your preferred rate and destination."

Additionally, on top of the thirty rupees per ride that Ola gives to the drivers, it offers an incentive for being logged on for ten hours. Interestingly, even with the per-ride and log-in incentives, the metered fare, and the ten rupees from the client, drivers still makes less than they would had they set the fare through haggling with the client. The difference is the phone calls they must make to clients and the extra gas used to reach them.

The drivers are aware that they do not make as much money per ride using Ola as they did in the old system. Why would they continue to work in a system in which they work longer hours but make the same amount as before? The drivers talked about wanting to keep up with the times, which they conflated with working more hours. They would rather work more hours for the same amount of money than sitting idly by, passing time drinking chai at the stand. As one driver remarked, "You have to change yourself with civilization/urbanization. Earlier, we used to rest a lot, but now we have to strain ourselves a bit. That is change."

Using trial and error, as the log-in incentives have decreased, most drivers have developed a method that combines catching rides from the stand or street with gaining clients through the Ola app. The sheer volume of clients on Ola makes it beneficial for drivers who need to find clients at a moment's notice. After dropping off a client at a remote location, they can find a new client at that location and then return to a more populous area. This system wastes less gas because the vehicle is always filled with paying clients.

IV: But, will you continue with Ola, or do you think you will give that up?

DRIVER: Nothing like that. Its okay. It's good. Whenever I am not getting a regular saavari client, then I will take an Ola call. Better than just sitting. And now, when I return, if I get a better saavari than what Ola has to offer, then I will take that up.

At the time of the interviews, CITU, the Communist Party of India's union, was warning drivers not to sign up with Ola. The union argued that fare structures should be set by the government and that Ola's ten-rupees-above-the-meter rule interferes in that process.[11] The drivers expressed distrust of the union. Across the board, their counterargument was that the unions had not fought hard enough to improve conditions for the drivers. Because they felt the unions were corrupt and complied with the government, they turned to Ola, a private enterprise. One driver remarked, "There actually shouldn't be an Ola. If the government enforces its rules and is consistent with its rules, we won't need an Ola. They should raise the meter rates when there is a petrol price hike. They will continue with the same meter rate for three or four years, when there is inflation of basic grain prices every six months. We won't have to ever depend on Ola then."

These interviews brought forth a few key observations. First, the drivers who had been drivers for most of their lives adapted to circumstances without abandoning earlier practices. They kept what worked for them and abandoned the rest, developing a hybrid way to meet clients. Second, Ola drivers do not have faith in traditional structures such as unions or the state to protect their interests and ensure good working conditions. Drivers had no expectation that the state would care for their well-being, and the union's lack of resolve and ability to keep the base metre rate in line with inflation has indeed left drivers in the lurch. Drivers have had no choice but to adapt to an economy in which productivity is valued over being compensated fairly. They would rather have peace of mind and take pride in their productivity than feel idle as they spend their days arguing with clients about fares.

References to technology are ubiquitous in India. Almost every song has a line about social media or computer hardware. Tech words have become indigenized and incorporated into the country's regional languages. The affective ties that workers have to technology cannot be underestimated. Within these songs, the allure of technology is its promise of social mobility and greater control over their lives. Workers imagine that working hard and working with technology is a step towards modernity, a step out of their current conditions. Technology coexists alongside traditional work practices. The analogue and digital labour come together in the everyday battle to make ends meet. In this creative "hacking" lies the everyday life and history of technology and labour in India.

NOTES

1 *Gram* is "village" in Tamil. *Instagram* turns into "Insta-village" in this context.

2 "Mersalaayitten," directed by A.R. Rahman, with Vikram and Amy Jackson, Sony Music, https://www.youtube.com/watch?v=uI_ug1H6u0k, 0:46–1:20.

3 Catherine Shu, "India Will Have 500 Million Internet Users By 2017, Says New Report," *TechCrunch*, July 21, 2015, https://techcrunch.com/2015/07/21/india-internet-growth/.

4 Wired connections through Digital India campaign.

5 Idea commercial, created by Lowe Lintas Agency, produced by Coconut Films, https://www.youtube.com/watch?v=TgUhZl9ZPVA.

6 Sainul K. Abudheen, "Ola Now Has $250–300M Annual Gross Transaction Run Rate," *VCCircle*, November 19, 2014, https://www.vccircle.com/ola-now-has-250-300m-annual-gross-transaction-run-rate-peek-its-numbers/.

7 Sounak Mitra and Digbijay Mishra. "Ola, Meru, TaxiForSure on Low Financial Mileage: Despite High Valuations, Most Taxi Service Providers Are Stressed," *Business Standard*, December 11, 2014, https://www.business-standard.com/article/companies/ola-meru-taxiforsure-run-at-low-mileage-114121000197_1.html.

8 Ashish K. Mishra, "Behind TaxiForSure's Sellout," *livemint*, March 20, 2015, https://www.livemint.com/Companies/t7TozTlZCAmvtSxog3OQ7L/Behind-TaxiForSures-sellout.html.

9 "Unrealistic Discounts, Driver Incentives: CCI Probes Ola for Predatory Pricing," *Firstpost*, May 7, 2015, https://www.firstpost.com/business/unrealistic-discounts-driver-incentives-cci-probes-ola-for-predatory-pricing-2231896.html.

10 Vivek Narayanan, "Auto Meters in Chennai Start Ticking," *The Hindu*, August 25, 2013.

11 Anannya Sarkar, "Unions Sore as Ola Woos Auto Drivers with Ride Incentives," *New Indian Express*, May 5, 2015.

The Most Hated Woman in Israel

Shaka McGlotten

I REMEMBER DRIVING away from Gush Dan, the greater Tel Aviv metropolis, rows of trees moving past in my peripheral vision. I remember the shared feeling of incredulity – five cops at Natali Cohen Vaxberg's door, one of them showing her his gun, trying to scare her. Amit and I disagreed about what she said. I told him, "One of the policemen said to her, 'We are the police,' and then he showed her his gun, just like in one of her plays. They didn't say they were thugs. But one said, 'We're police, and [we're] criminal' and 'you're a criminal, and we'll treat you like one.' I'm almost a hundred percent sure." I remember that she said that she had shouted for one of her neighbours. She was charged with resisting arrest and assaulting a police officer. She didn't have socks on, so the cuffs dug into her legs. One cop questioned whether this was really necessary, but she stayed cuffed.

When Amit and I visited her days later, bruises were still visible on her arms.

She'd been arrested for her video "Shit Instead of Blood," released online in the wake of Operation Protective Edge, the 2014 assault on Gaza.[1] The actual charges were "hurting the public's feelings" and "defiling national symbols." When I wrote to Israeli legal scholar Professor Aeyal Gross about the case, he told me that the former had traditionally been reserved for settlers who caricatured Muhammad as a pig.

It was only by listening to the debrief my then-partner Amit and I had recorded in the car after the arrest that I was able to reconstruct the conversation we had that night and where we were headed – to his dad's in Be'er Sheva, where none of this, not the assault or Natali, would be discussed. Sustained conversation would only lead to an ugly family fight. Instead, we spent our evening eating an elaborate meal prepared by his stepmom, drinking much-hyped (and pretty decent) Israeli wine, and planning the next day's visit to the local strip mall.

"Not the exit to Ashdot, take the next one," Amit instructed.

We zipped along. The highways in Israel are well built. Maybe we even took the new toll road.

I remember an argument, or our usual arm-wrestling, but it's not in the recording. I can tell, though, listening to the recording. I'm tight-voiced, persistent, "No, let's really try to reconstruct what she said. If we can get a hook into it, we can unravel the whole conversation."

I was pushing Amit to try to remember key words to get my memory jump-started. Is this what ethnography looks like? I suppose so, in spite of my ongoing imposter syndrome as an anthropologist and the sense that my ethnographic work has only ever failed.

As I first drafted this chapter in 2016, two years after our visit, I had to use Google Maps to look up exurbs of Tel Aviv – I couldn't remember the name of the city where Natali lived. It's Holon, coincidentally the sister city to Berlin's central borough Mitte, immediately adjacent to Wedding, where Amit and I lived in 2014 while I was conducting research on Berlin's drag scenes; Israel was another site for that research. Natali's suburb struck me as a little sad and run down, but a lot of Israel feels like that; it was modern once; it was never modern.

The nondescript apartment building Natali was in – she shared the flat with her mom – was only four or five storeys tall. It exudes a working-class vibe, but it's most likely inhabited by all of the middle-class people who have been pushed out of Tel Aviv. But the skyrocketing rents in Tel Aviv combined with development efforts mean it's gentrifying, too. Officials have rebranded Holon as a children's city: bring your kids, have kids, go to the Children's Museum. The streets were so empty when we visited that I thought it must have been Shabbat, but when I checked our shared, still shared, post-divorce calendar from 2014 for confirmation, I learned that we had visited her on a Tuesday.

Somehow it hadn't felt right to record Natali, even though we were there for an interview. Still, the anthropologist leaned in, paying attention to everything: the modest space, her mom's old paintings, Natali's embarrassment at her English. Natali seemed fragile – she's so tiny, and the bruises were an ugly purple against her skin. Yet I didn't think of her as a victim either. Her YouTube videos kept coming, after all, one after the other, even after the prosecutor declared that she'd be made an example of.[2] I also didn't think of her as a victim because she was not sorry, not for anything at all.

She might be "the most hated woman in Israel," as she was named during her first television interview, temporarily supplanting parliamentarian Hanin Zoabi, but her lack of remorse, as well as the tea and cookies she'd laid out for us, endeared her to me from the very beginning.

Natali has a big mouth. I mean, her mouth is actually very big, and she is also a big-mouthed troublemaker, shouting sense in the form of absurd theatrics to her fellow Israeli Jews. Although Israelis are stereotyped as being pushy and loud, these qualities are also vital elements in Israeli theatre histories. Hanoch Levin, a playwright known for his often vulgar satire, is a beloved icon now that he is dead, but when he was alive he bitterly divided the public. His work, like Natali's, was frequently scatological and frequently censored. His controversial play *Queen of the Bathtub*, which took aim at Golda Meir, featured a character called Lord Keeper of the Enema.

Listening to an interview we had conducted with Natali via Skype, earlier in 2014, I hear how Amit and I laughed when she described an audience's reactions to one of her plays, performed at the Acre theatre festival. When people walked out and shouted at the performers, she was delighted. By that time in our relationship, Amit's voice and mine had taken on the same timbre. In the recording, I sound a bit like a slightly more animated than usual National Public Radio reporter, but with an international flair – I'd gotten used to speaking English with non-native speakers. Amit, a mimic, had inherited my cadences, but I hear him fall out of them when he shifts to Hebrew to translate something for Natali. Listening, I can't tell which of us laughed when she recounted the story of being censored by a German festival. The organizers had invited her to perform her Shoah work, in which she personifies the Holocaust, and then told her after she'd arrived that they had changed their minds. They were afraid to be seen as anti-Semitic, of course. When the organizer who'd invited her stopped returning her emails, she took a photo of herself at Thomas Mann Strasse, holding a sign that read "I'm proud to be on the list of Jewish writers that were censored by the Germans." (Uncannily, Amit had a similar experience a couple of years later. His critique of German antiblack racism was mistaken for the thing itself by the Germans who'd invited him to participate in a series of performances in former synagogues around the country. He altered the piece out of respect for the performers.)

In her performances, Natali's big mouth issues forth a mythical Voice, alternatively hysterical (in the best sense), childish, and scolding. It takes her fifteen minutes to put on the Voice's costume – a bold slash of red lipstick. It takes her so long because it's not for daily life; it's something special, something draggy, like the poop tiara she wore when she appeared on TV to discuss her arrest. I tried it on and took selfies when we were at her apartment in 2014. So her mouth got her in trouble, and her ass did too.

Her arrest had gotten in the way of our original date. And communicating was hard. Amit and I were already in Israel but had to coordinate the interview through a friend. Natali's means of communicating with the outside world had been taken away. Israel prides itself on being a high-tech country, and it has produced a number of innovations, although a little digging reveals the ways they are almost always tied to its military-industrial complex (what isn't?). So there's Wi-Fi everywhere, and Israelis are as glued to their smartphones and Facebook and YouTube as most of my fellow Americans.

After a night in jail, there were a couple of days of house arrest, but most importantly Natali was banned from using the internet for a month, and she'd had her phone and computer confiscated for three months. After that second Acre festival play, the one where people walked out, just like with some of Levin's plays, Natali had essentially been blacklisted by the Israeli theatre world. The web had become her main medium. "I couldn't get my work produced, but, hey, I thought, there's YouTube." Of course, the hecklers followed her there, but their criticisms now took the form of views, likes and dislikes, and often vicious trolling. No phone, no computer: she was being punished like a child on a time out.

For what? For pooping on forty flags, including Israel's.

It was Yoav Eliasi, a.k.a. the Shadow, a right-wing rapper turned media pundit who had narced on her, directing authorities to the video, the same Shadow who, in response to that summer's (small, modest) protests against the Gaza assault, had used his Facebook page to incite violence against anyone critical of the murderous attacks, that is, to beat up the real enemies of Israel: leftists and Arab citizens. Some of the right-wing protesters wore neo-Nazi T-shirts. Police often stood by while this happened, watching. Later, a friend told me that during one confrontation in Tel Aviv that summer, an air-raid siren sounded, and all of them – leftists, right-wing mob, and police alike – huddled in a shelter together until the siren's wail quieted.

It took Natali three months to make that video, and it's not only her doing the pooping. Her friend and collaborator Jasmin Wagner helped; so did her dog, which she managed, after much effort, to film pooping on a world flag. Shit on all nations, all borders! She told us that because pooping isn't always easy for her, she used suppositories, a recurring motif in her plays. In *Only Death Will Discharge Us from the Ranks*, glycerin suppositories are characters. (The stage directions read: "The glycerin suppository will appear dressed in an elastic outfit, its hands kept close to its body. The glycerin suppository will have a snorkel attached to its head, and even some sort of crown – a candle like at a Chanukah party. Maybe even some glasses with exaggeratedly big eyes drawn on them.") I gave her some unsolicited advice about pooping from my Ayurveda training. Eat cooked food. Sweet potatoes are good for you. Your poop should be the consistency of a smoothie. Amit rolled his eyes. He'd heard it all before.

Even before she was arrested, she described her poop events as crime scenes. She tried to time her dumps to when her mom was away, committing little crimes and then disposing of the evidence before she returned.

In 2013, Cohen Vaxberg had herself gone to the police. She had a stack of death threats in response to another performance, one in which she performs as the Shoah, the Holocaust.[3] The Holocaust project began when she wondered what the Holocaust would feel if it were a person. She explained to us: "And the first thing was that they treat her like a whore. She is a celebrity and a bit of a whore. She needs the crowd." In the performance she did at Yad Veshem, the Holocaust memorial in Jerusalem, a site designed to induce sober memorialization, Natali, as the Holocaust, comes across like a deranged kindergarten teacher: sweet, sly, vicious. Flanked by friends who pose/serve as security, she is dressed in a slip, a tiara crowned her hair. She excoriates the witnessing public but especially politicians who regularly seek to benefit politically from the Jewish genocide. She might seem a little hysterical, the Holocaust, but Jerusalem is, of course, a hysterical city. I'm a softie, an anthropologist, a spiritual practitioner, but on my first visit there in 2010, I understood why nations such as the Soviet Union might seek to ban religion.

It is thanks to me that you can develop an atomic bomb and no one objects. For that alone I deserve thunderous applause!

Without me ... you couldn't run a ghetto for 3 million people!

I'm the best thing that ever happened to you!

Natali made a video addressed to Palestinians too: "Hello Palestinians. I am the Holocaust. Because of me you had the Nakba (catastrophe)."[4] Of course, the Holocaust didn't cause the Nakba – that catastrophe was predicated on histories of settler colonialism – but you wouldn't know that because the dispossession of Palestinians is so often justified by the need for a Jewish state in the wake of the genocide.

The *Holocaust's Visit to Yad Veshem* has more than four hundred thousand views. The comments are illustrative, an unsurprising mix of the best and worst YouTube comment threads have to offer. Some are supportive – "Great video. Keep up the good work." – while others express misogynist or anti-Semitic views. Plenty of the comments are by trolls who created an account to say something trollish. Some of them are probably Jewish students being paid to troll anti-Zionist sentiment.[5]

The police hadn't done a thing about the death threats against her. But her shit was troublesome enough to warrant her arrest and continued intimidation by the state. Rogel Alpher notes in a *Ha'aretz* editorial that Cohen Vaxberg "belongs to a glorious and even pioneering Israeli tradition of using feces to express grievance."[6] He points readers to a 2009 report by the human-rights group B'Tselem, which documents abuses under the Israeli occupation. The report, titled *Foul Play*, reveals how wastewater from Jewish settlements in the West Bank is allowed to flow untreated into Palestinian waterways and farmland.[7] Copropower, the political operationalization of shit, is an important tool of biopolitical management in Israel/Palestine, as elsewhere.[8]

In the silences during our drive south, we worried about Natali. She had told us that she felt isolated and was afraid to go out. Friends were getting groceries for her. Amit pointed out an especially ironic twist to the story. Like most artists living in societies such as Israel's – neoliberal, deeply unequal – she had to hustle to make ends meet. When we met with her, her day job was, in an irony she readily acknowledged, dog walking – walking other people's dogs and picking up their shit. Her new fame could jeopardize that. Amit imagined a scenario: the neighbour who won't use Natali's services anymore: "She can't pick up the poop of my dog. She poops. She's crazy."

"Crazy," that's a word that gets applied to Natali a lot, like the word "extreme" that Amit came to proudly own, since in Israel, "extreme" means supporting Palestinians' right to determine their own future. Amit continued, listing the names of some of our friends who are deeply involved in pro-Palestinian causes, from people involved in front-line protests against the Apartheid Wall to academics, artists, and legal activists. They called her crazy too. Left-wing political blogs attacked her, afraid of being associated with her shitty leftism. And the art world? Totally silent.

In 2014, Prime Minister Benjamin Netanyahu suggested stripping citizenship from Israelis who called for the destruction of the state of Israel, by which he meant artists such as Natali or leftists (even the cowards who won't stand up for her) or Israeli Palestinians who question whether a "Jewish democracy" can ever manage to be anything other than an ethnocracy.

Before too long, Natali had posted another video to YouTube in response to her arrest and Bibi's nationalistic threats.[9] This one mimicked the aesthetics of ISIS beheading videos. In it, she kneels in an orange jumpsuit; a man with a knife stands to her side. She begins: "I'm sorry ... that my shit isn't blue and white, blue and circumcised, more round, less triangular ... blind to flags and symbols is my poo."

NOTES

This essay is part of a larger project about drag and politics and was supported in part by the Alexander von Humboldt Foundation and Akademie Schloss Solitude. I thank the editors for their invitation to contribute, Natali for her generosity, and Amit Gilutz for providing feedback.

1 Natali Cohen Vaxberg, *Shit Instead of Blood*, video, 2014, https://vimeo.com/100127568.
2 Yaniv Kubovich, "Police Expected to Indict Artist for Defecating on Israeli Flag," *Ha'aretz*, January 13, 2016.
3 Natali Cohen Vaxberg, *The Holocaust's Visit to Yad Veshem*, video, 2014, https://www.youtube.com/watch?v=flfUvPyLVZI.
4 Natali Cohen Vaxberg, *How Would You Manage without the Holocaust?*, video, 2012, https://www.youtube.com/watch?v=gPLGczT6Hjw.
5 Ali Abunimah, "Israeli Students to Get $2000 to Spread Propaganda on Facebook," *Electronic Intifada* (blog), January 4, 2012, https://electronic

intifada.net/blogs/ali-abunimah/israelistudents-get-2000-spread-state
-propaganda-facebook.

6   Roger Alpher, "Where Are the Indictments against Those Who Use Poop to
    Oppress?," *Ha'aretz,* January 16, 2016.

7   Eyal Hareuveni, *Foul Play: Neglect of Wastewater Treatment in the West Bank,*
    report for B'Tselem, June 2009, https://www.btselem.org/download/200906
    _foul_play_eng.pdf.

8   Shaka McGlotten and Scot Webel, "Poop Worlds: Material Culture and
    Copropower (or, Toward a Shitty Turn)," *Scholar and Feminist Online,* 13.3–
    14.1 (2016), http://sfonline.barnard.edu/traversing-technologies/poop-worlds
    -material-culture-and-copropower-shaka-mcglotten-scott-webel/.

9   Natali Cohen Vaxberg, *A Video in Response to the Shit-on-a-Flag and Arrest,*
    video, 2014, https://www.youtube.com/watch?v=7RJyRv5SzMU.

DIY WI-FI

Heather Frise

SINCE IT WAS FIRST conceived in the early 1970s, the internet has evolved from a modern luxury, to a household necessity, to – according to a 2011 UN resolution – a fundamental human right.[1] In July 2016, the UN elaborated on this resolution, declaring the importance of "applying a comprehensive human rights–based approach when providing and expanding access to the internet and for the internet to be open, accessible and nurtured."[2]

Although this resolution acknowledges the need for the universal recognition of digital rights, billions of people around the globe still struggle against major economic, geographic, and political barriers to unfettered internet access. In response, citizens worldwide are experimenting with low-cost technologies in an attempt to create affordable, alternative internet networks, thereby eclipsing corporate and government control.

One of the biggest autonomous internet networks in Europe began as an effort to connect isolated areas and islands in Greece that weren't being served by telecom companies. "At first, back in 2002, it sounded like a crazy idea," IT expert Joseph Bonicioli said in an interview for the National Film Board of Canada's *Highrise* project. "It started mainly due to the lack of broadband, and as a technological experiment to see that people could share their connectivity."[3] Bonicioli paid a service provider a monthly fee to get backbone access and then shared the connection using wireless rooftop antennas.[4] The Wi-Fi signal passed from one antenna or node to another, and one by one a chain of nodes formed a mesh. The network was open to anyone who wanted to join and share their infrastructure.

This grassroots mesh has grown into a vast and vibrant online community called Athens Wireless Metropolitan Network (AWMN). AWMN has more than three thousand members that range from Athens to the nearby islands. Data moves through the AWMN mesh up to thirty times faster than it does on the telecom-provided internet, and members have developed their own services for classified ads and blogs as well as an internal search engine. "AWMN is our own personal network," said Bonicioli in the interview, "we formulate it the way we want it; it's our personal playground." Bonicioli highlights the social aspect of the network: "This is not just a mesh of equipment, it is a mesh of people. It's a network of people. People have created small communities in different areas of Athens. Great relationships have formed. People have become friends. Some have even married." Today, the network is still completely free, aside from the cost of the Wi-Fi hardware required to set up a node.

Bonicioli's DIY approach has been echoed in numerous initiatives around the world. The community network Guifi began in the early 2000s as a way to bring the internet to rural Catalonia, where commercial service providers weren't offering broadband access. Guifi cofounder Ramon Roca initially gained access by attaching long-range antennas to Wi-Fi cards and pointing them at public hotspots such as libraries. Similar to AWMN, in this network his connection could be shared with anyone so long as they set up their own networking equipment. Currently, Guifi has more than thirty thousand network links, the majority of them concentrated in Catalonia along Spain's Mediterranean coast, though the mesh is as far-reaching as Argentina, China, India, Nigeria, Yemen and, mostly recently, the United States.[5]

In Argentina, an organization called Altermundi formed to bring free internet connection to small towns and villages where no commercial service was available. A mobile team of digital activists travels to communities, often at the request of an individual or an NGO, to construct a mesh network using free software and Wi-Fi routers. Altermundi provides the technology and the training, but it is the community's responsibility to maintain the network.[6]

Other innovations include the creation of mesh potatoes – small, inexpensive Wi-Fi devices that use open-source technology to mesh automatically with one another, allowing people to transmit data and make local calls. Mesh potatoes were designed by Stephen Song of

Village Telco and first used in towns across Africa, where internet access is overpriced or nonexistent. Local shops buy backbone access and then sell mesh potatoes to customers, offering them cheap monthly phone and internet rates.[7]

In Ghana, Accra high-rise resident John Ampah started up a Wi-Fi business. "It wasn't for money," he said, "I just had the need to help people who needed the internet." Using long-range antennas, he shared his Wi-Fi signal with over one hundred households. People who would otherwise not have the means can access and participate in digital life: a father can use the internet to do math homework with his children, a seamstress can download dress patterns, and a welder can buy tools and materials for his business.[8]

These grassroots networks have the advantage of being private, protecting users from corporate and government surveillance. Privacy is critical for all citizens in the post-Snowdon world, but it is especially important for social and political activists. Although it's true that no network is completely private so long as it is dependent on backbone signals or satellite connections, in these networks it is much more difficult for corporations and governments to track online activity or gain access to personal data.

The New America Foundation's Open Technology Institute, a nonpartisan think tank in Washington, DC, has developed Commotion – a mobile DIY kit containing software that allows anyone, anywhere, to quickly set up a mesh network. The software could be used during a protest or in the wake of a natural disaster, as it was after Hurricane Sandy in Red Hook, New York. The ability to control autonomous networks is especially crucial in countries with repressive regimes. In a country such as Egypt, for instance, during the 2011 uprising, social media was an accelerant for the movement. When President Mubarek shut the internet down in an attempt to contain and censor protesters, autonomous networks allowed activists to continue organizing and to garner international support. The widespread use of social media sent a message of hope to other activists in the Middle East and beyond.[9]

Joseph Bonicioli also put his mesh network to use in a political context. In 2011, during the Occupy movement in Athens, daily rallies against austerity measures took place in Syntagma Square. When the government shut down the internet, Bonicioli and his team responded by setting up their own tent in the square, building a node, and connecting thousands of protesters to the AMWN network.

Three years later, when the government pulled the plug on the Hellenic Broadcast Corporation, the state-owned public broadcast station, thousands of citizens protested in the streets. Journalists occupied station headquarters after their eviction, prompting a police raid and a government order to shut down the internet. Bonicioli and his tech team responded, once again, by rigging up a local node. Their actions allowed the journalists to continue broadcasting from inside the station, transmitting their message to the world.

The digital revolution is revolutionary only in as much as it opens up truly free and democratic tools of expression and collective digital organization. So long as governments and corporations abuse their power and erode civil liberties, the need for more independent, grassroots alternatives to internet access will continue to grow.[10] At the same time, it will continue to be important to hold government and corporations accountable for their infringements on privacy. Doing so will require new pathways for protecting digital rights and sweeping changes to surveillance laws.[11]

In 2013, Clive Thompson, who writes about the social and cultural impact of technologies, speculated about the implications of "a mesh that could span the entire globe, a way to communicate with anyone, anywhere without going over a single inch of corporate or government cable."[12] Indeed, this is already starting to happen as wireless communities worldwide start linking their networks together. "Ultimately," says Bonicioli, "we believe in the participation of people, the creativity and the building of local communities that will make the future of networks more democratic and humane."[13]

But the creation of autonomous networks is not inherently neutral or democratic. Deliberate measures need to be taken to prevent the reproduction of inequities and existing power structures and to ensure that substantive roles and decision-making power are available to everyone, especially those most affected by the digital divide. Meaningful internet access for women, for instance, remains elusive. According to the International Telecommunication Union's latest research, women are 12 percent less likely than men to have access to the internet; in the least developed countries, that figure jumps to 33 percent.[14]

On a practical level, it's difficult for some, especially more remote, communities, to maintain network infrastructure and keep equipment running smoothly. And as telecom connectivity expands across the globe, community networks have to compete with the ubiquity of smartphones. People tend to favour the apps and internet access afforded by their phones over reliance on and participation in community networks.

Many communities in remote areas want to be able to access global networks if they are available. And while connectivity has a plethora of social, political, and economic advantages, it also brings with it invasions by commercial and, often predatory, forces. Connectivity runs the risk of turning empowered citizens into passive consumers, supplanting local culture and diversity for global monoculture. It also makes users more subject to punitive data collection and data-sharing practices by police, welfare offices, hospitals, immigration agencies, and other institutions.

Balancing access and inclusion against the centralizing forces of the internet is an ongoing negotiation. And yet, half the world's population still does not have access to this basic human right. Autonomous networks can both bridge the digital divide and foster new internet paradigms that are fair, trustworthy, and truly collaborative, paradigms or living systems that are as dynamic and complex as the communities that create them.

NOTES

1 Frank La Rue, "Report of the Special Rapporteur on the Promotion and Protection of the Right to Freedom of Opinion and Expression," May 16, 2011, A/HRC/17/27, https://undocs.org/en/A/HRC/17/27.
2 Tim Sandle, "UN Thinks Internet Is a Human Right," *Business Insider,* July 22, 2016, https://www.businessinsider.com/un-says-internet-access-is-a-human-right-2016-7.
3 Joseph Bonicioli, interview by Heather Frise for the National Film Board documentary *Highrise,* January 17, 2015.
4 Michael Kende, "The Digital Handshake: Connecting Internet Backbones," *CommLaw Conspectus* 11 (2000): 45–70.
5 L. Finch "guifi.net, Spain's Wildly Successful DIY Wireless Network," *Rising Voices* (blog), December 11, 2013, https://rising.globalvoices.org/blog/2013/12/11/guifi-net-spains-wildly-successful-diy-wireless-network/.
6 Leila Nachawati Rego, "AlterMundi: Community Networks Embody the Original Spirit of the Internet," Association for Progressive Communications, news release, November 23, 2015, https://www.apc.org/en/news/altermundi

-%E2%80%9Ccommunity-networks-embody-original-spirit-internet%
E2%80%9D.

7   David Rowe, "The Mesh Potato," *Linux Journal,* December 1, 2009, http://
www.linuxjournal.com/magazine/mesh-potato.

8   John Ampah, interview with Jacky Habib for the National Film Board
documentary *Highrise,* March 20, 2014.

9   Sam Guston, "Social Media Sparked, Accelerated Egypt's Revolutionary
Fire," *Wired,* February 11, 2011, https://www.wired.com/2011/02/egypts
-revolutionary-fire/.

10 Ewen Macaskill and Gabriel Dance, "NSA File Decoded: What the Revelations Mean for You," *The Guardian,* November 1, 2013.

11 Matthew Taylor and Nick Hopkins, "World's Leading Authors: State Surveillance of Personal Data Is Theft," *The Guardian,* December 10, 2013.

12 Clive Thompson, "How to Keep the NSA Out of Your Computer," *Mother
Jones,* September–October (2013), https://www.motherjones.com/politics/
2013/08/mesh-internet-privacy-nsa-isp/.

13 Bonicioli interview.

14 Gabrielle Willms, "Helping Women and Non-binary Communities Gain
Access: Reflections from the Best Practices Forum on Gender and Access
on the Potential of Alternative Models of Connectivity," *APC News,* March
7, 2019, https://www.apc.org/en/news/helping-women-and-non-binary
-communities-gain-access-reflections-best-practice-forum-gender-and.

● Network Dislocations

Nicole Starosielski

A DENSE TOPOGRAPHY of wires sprawls beneath the city surface, circulating messages through basements, conduits, and tunnels. Aerial cables connect telephone poles and buildings. Wireless signals move from mobile phones to cell towers and ultimately onto the fibre-optic network. Traffic transits between networks at internet exchange points and accumulates on corporate computers. The urban accretion of digital infrastructure is evidence that infrastructure attracts infrastructure.

The submarine cables that funnel signal traffic across oceans and connect cities into a global network are largely absent from this landscape. Even though the traffic that they carry is often destined for urban hubs, underwater cables typically come ashore in rural and suburban locales. For these cables, deemed critical infrastructure essential to national security, the density of urban traffic, whether human or technical, has been a threat. Cities both attract infrastructure and repel it.

Here, I track the movements of cables towards and away from urban hubs around the Pacific Ocean, revealing the insular nature of cable geographies – their need for shelter from points of interconnection. The images, which are themselves composed of layers of cable lines, text, and photography, are drawn from *Surfacing* (Surfacing.in), a digital project that makes visible the hidden terrain of the undersea network.

When the first telegraph networks were strung around the Pacific, they were set up at both urban locales and rural outposts. By geographic necessity, some cables landed in remote towns, including Darwin, Australia; Cable Bay, New Zealand; Bamfield, Canada; and Cape St. James, Vietnam. Others were linked directly to city centres.

San Francisco was the most important city to trans-Pacific telegraphy in the Americas. It was an ideal location to interconnect with domestic

San Francisco, California. Courtesy surfacing.in/?map=sanfrancisco.

lines: east of the city, the Overland Telegraph extended to New York; to the south, signals flowed down the coast to Santa Barbara, Los Angeles, and San Diego. Across the ocean, a telegraph cable linked San Francisco to stations in Manila, Hong Kong, and Singapore, enmeshing it in an electrically networked colonial geography.

When undersea telephone cables replaced telegraph lines in the 1950s and 1960s, many of the original urban landings were abandoned. In California, cables were relocated from San Francisco to rural Manchester and San Luis Obispo. A similar shift occurred in the Philippines, where a telegraph station in Manila was replaced by a link to Nasugbu, a small coastal town sixty miles from the capital. And while Tokyo's traffic once filtered through Yokohama, the postwar submarine telephone system landed in suburban Ninomiya, more than forty miles outside Tokyo. It was technically feasible for many of these systems to directly connect to city centres, but the buffer between remote cable station and urban hub formed a critical barrier of protection – from nuclear attack, nearby populations, and the anchors of local boats.

Although cables were moved far beyond city boundaries in California, Manila, and Tokyo, in Hong Kong the shift was much less pronounced.

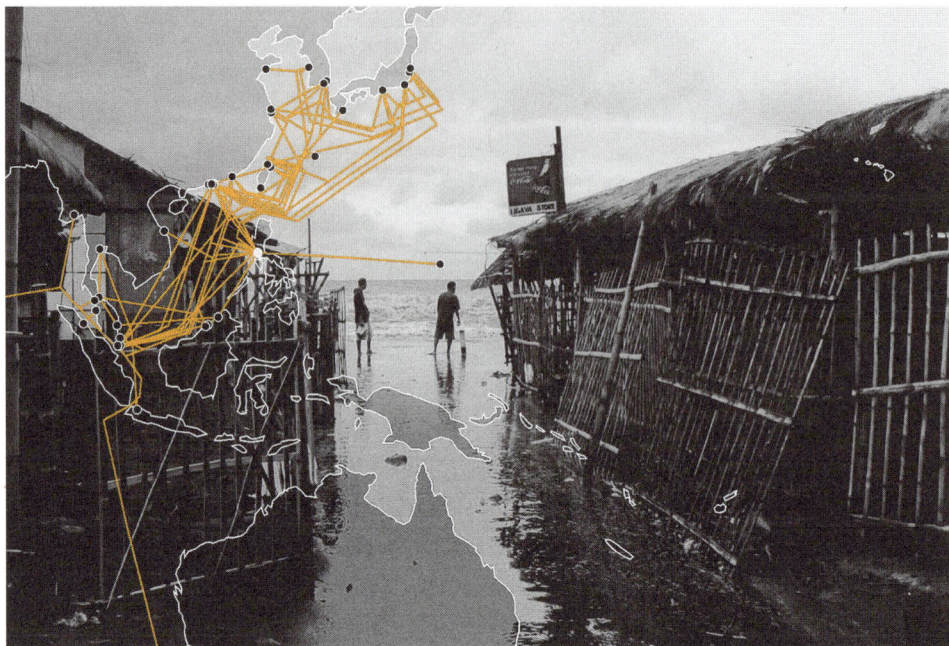

▲ **Nasugbu, Philippines.** Courtesy surfacing.in/?map=nasugbu.

▼ **Lantau Island, Hong Kong.** Courtesy surfacing.in/?map=lantau -island-hong-kong.

▶ **La Perouse cable station, Australia.** Courtesy surfacing.in/ ?map=botanybay.

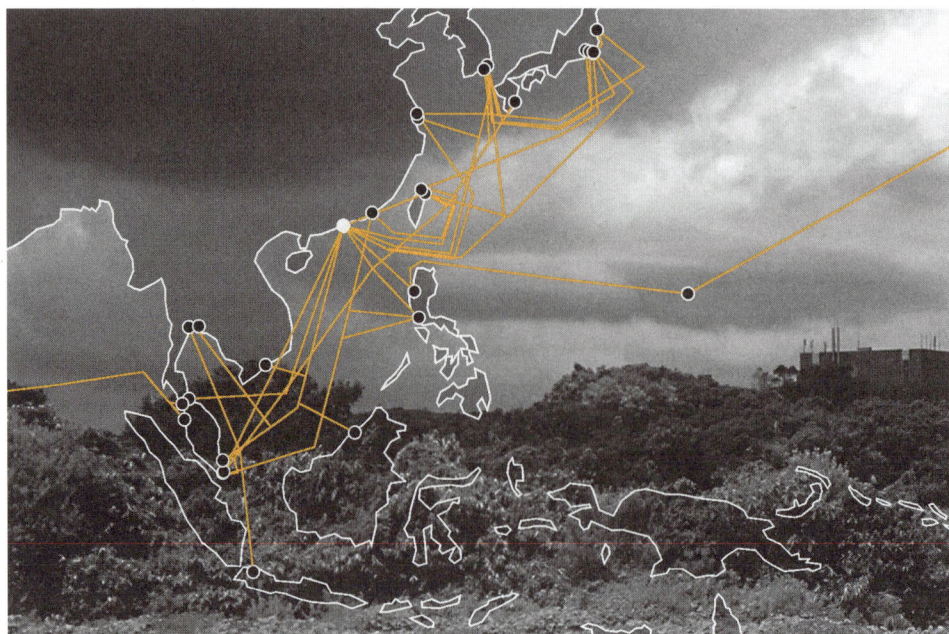

Over time, cable stations have slowly migrated, along with other industries, to suburban zones. The earliest telegraph cables to Hong Kong were laid to Telegraph Bay and connected to offices in the city's centre. With the establishment of the undersea telephone system, a new cable station was built on the island's less populated southern shore, at Deep Water Bay.

Over the years, the hills surrounding Deep Water Bay became high-end real estate and were populated by Hong Kong's elite. In the development of subsequent networks, telecommunications companies could no longer afford to build stations there. They moved farther away from the previous landings to Cape D'Aguilar and Chung Hom Kok (at the eastern edge of Hong Kong Island), to Tseung Kwan O on the mainland, and to Lantau Island, which had recently become home to Hong Kong's new international airport. The increasing distance between networks and the urban centre was not simply a safety measure to prevent physical disruption but a response to fluctuations in real estate and the industrialization of the urban edge.

There are some anomalies in this pattern. In 1876, the first submarine cable to Australia's east coast came ashore at Botany Bay, about ten miles south of central Sydney. Like many zones of cable activity, the La Perouse headland was a well-trodden landing point. British and French explorers had anchored there a century before, and in the intervening years, colonial settlers fortified the area, constructing a watchtower

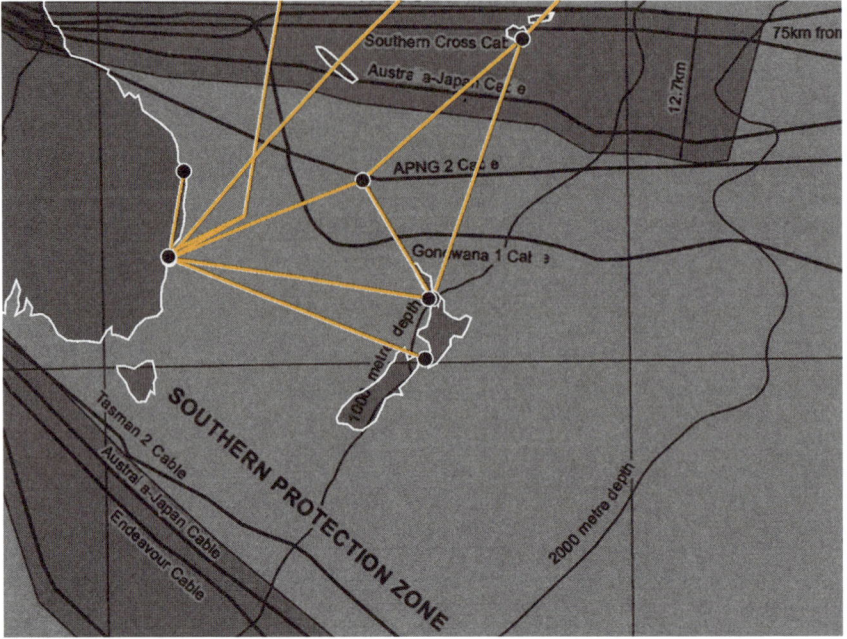

▲ **Bondi Beach cable landing point, Australia.** Courtesy surfacing.in/?map=centralsydney.

▼ **Vung Tau cable landing point, Vietnam.** Courtesy surfacing.in/?map=vungtau.

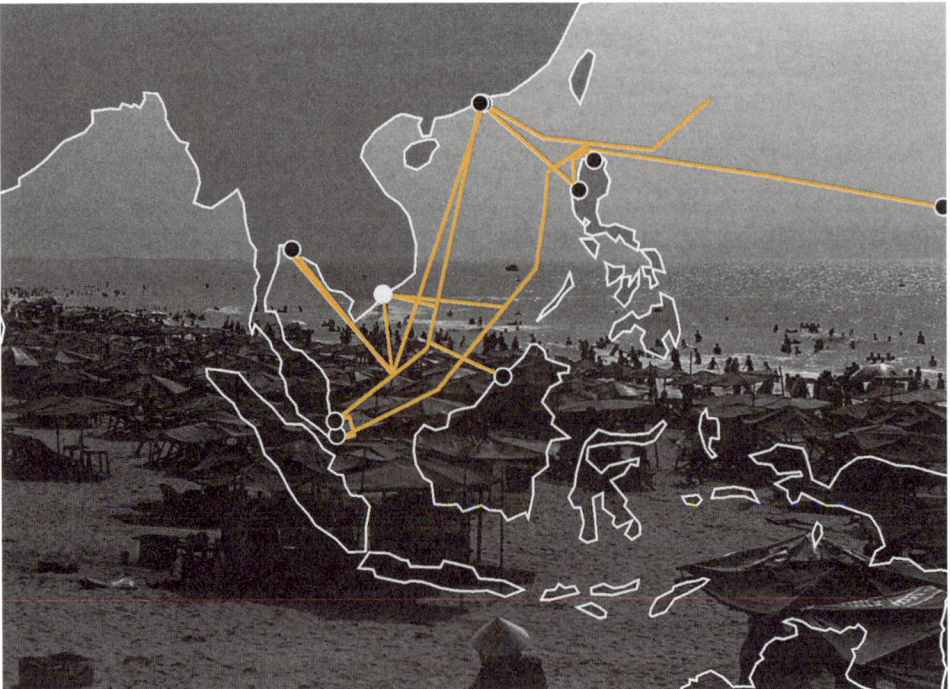

specifically to prevent smuggling. After a fire broke out in the station, cables were redirected north to Sydney's Bondi Beach, and from 1917 onwards, telecommunications were routed directly into the city, where the urban environment exhibited a gravitational pull on later installations. When the country's first undersea telephone link, the Commonwealth Pacific Cable, was laid in 1963, it, too, landed at Bondi Beach and from there was routed underground to Paddington Station.

Even at Bondi Beach, builders considered moving the station and returning to La Perouse, or to another less central, and thus less vulnerable and less expensive, location. But it wasn't feasible to shift the route away from the existing conduits and support infrastructure. Today, a network once centralized at Paddington Station has given way to an array of suburban nodes. Sydney is home to five different submarine landing stations, more than the rest of the country has combined. Internet traffic is concentrated there in part because of two cable protection zones that prohibit anchoring near network lines. These zones protect cable systems from other interurban circulations, including large tankers and container ships, but such consolidation, even at the periphery, makes the system as a whole susceptible to disruption.

In other places, once-remote cable hubs became cities themselves. During the French colonization of Vietnam, telegraph cables were landed at Cape St. James, at the mouth of the Saigon River. About sixty miles outside the city, cable operators were in "danger of interruption from the incursions of wild beasts," and a large tiger was reportedly killed on the verandah of the cable station. Today, the cable station is only a few miles inland from its original location but is now surrounded by Vung Tau, a city populated by almost half a million people.

Contemporary undersea cable stations are often located at the urban fringe, amid industrial areas, technology parks, and small beach towns. These displacements buffer the network from the movements of boats and the backhoes of local construction crews, from parallel wires in crowded subterranean conduits and the men who maintain them. This topography is a form of insulation that functions not unlike the cable's polyethylene covering – it ensures that signal traffic will flow unimpeded into the thick circulations of the city.

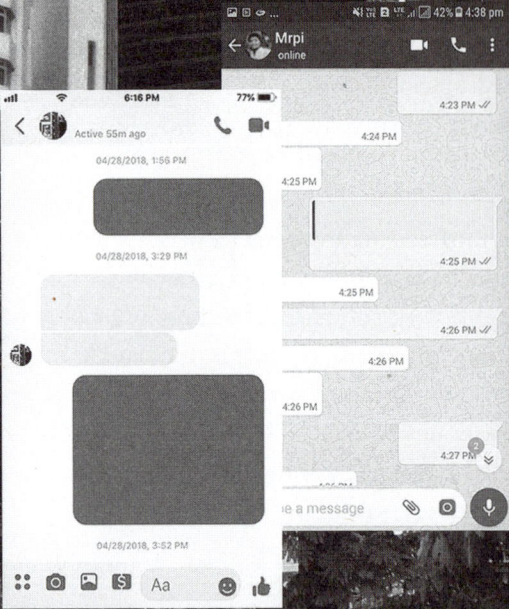

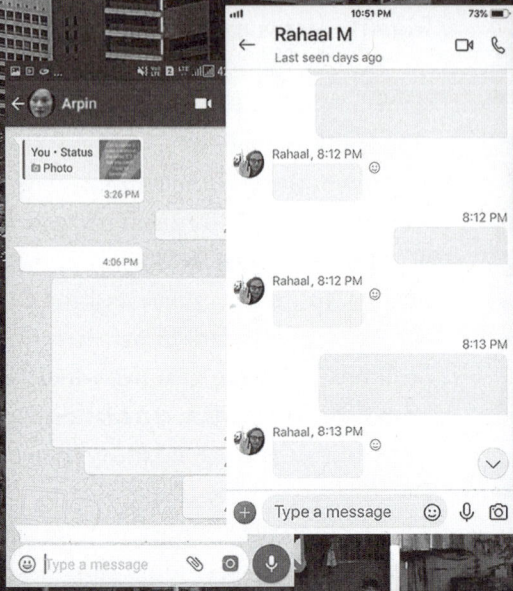

SINGAPORE

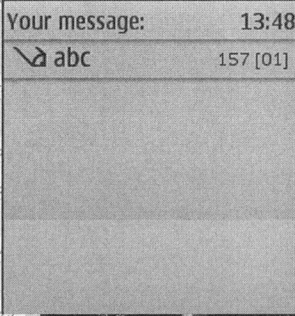

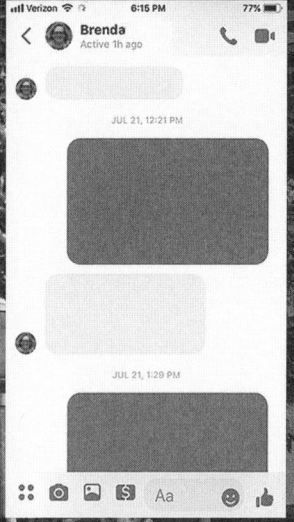

SINGAPORE

The Labour of Global City Building

Alexis Mitchell, Deborah Cowen

NOT EVERY COUNTRY showcases its national public-housing system in a museum, but Singapore is not just any country. Public housing is a linchpin of the national economy and even national identity in the city-state, and the Housing Development Board (HDB) is the agency charged with its construction, marketing, sales, and management. Today, more than 80 percent of Singaporean citizens live in public housing built by the HDB, the public body established in 1960 to tackle the postcolonial housing problem. It is perhaps not surprising, then, that the HDB Gallery proffers a vision of public housing that blurs the private and national meanings of "home."

The HDB Gallery is part theme park, part sales centre, and part museum. It takes visitors on a high-tech, interactive, digital experience that shows them how both the private family and the nation-state have and will continue to prosper from financialized public high-rise development. Before visitors even enter the gallery, they are met with scaled models of current HDB high-rise projects in which Singaporean citizens are encouraged to invest. Upon entering the gallery, visitors are ushered into a circular, windowless, spaceship-like room. As the door slides closed, they're surrounded by a 360-degree panoramic theatre, and an immersive experience of residents' lifestyles plays to the tune of Singapore's national song, "Home." The sixteen-minute video is profoundly celebratory and geared towards selling both the concept of a state-run housing system and the individual housing units within.[1]

The remainder of the exhibit also makes extensive use of digital technologies – including a "first-of-its-kind life-sized hologram," an "exercise-wheel activated video show," a series of "multi-touch interactive displays," and an "interactive game" – as the privileged medium for marketing this national and architectural vision of home.[2] Through these inter-active technologies, visitors are invited to meet multicultural, hetero-sexual Singaporean families. Holographic videos are projected onto the cupboards of model kitchens, where Chinese, Indian, and Malay couples explain how their HDB flats have enabled their families to grow and thrive. The focus on the family is ubiquitous in the gallery but also in the policies that sculpt access to public-housing units in the city-state that surrounds it. As the HDB's website explains, residents must "form a proper family nucleus," strictly defined in procreative heterosexual terms, to access units. It is not just the housing units that are being modelled; visitors are also invited to inhabit the normative familial citizenship of the smiling couples.

If the gallery highlights the seemingly magical powers of the family in building the nation, in and through the home, it entirely avoids men-tion of the lives and labour of those who actually *construct* and maintain the national housing infrastructure. This is particularly striking in a state that relies on a notoriously unequal system of migrant labour sup-ply. Elsewhere in this book, in a discussion of terraforming and the sand economy, Charmaine Chua highlights how Singapore is increasingly be-ing assembled through the theft of the raw materials that make up the land. Inevitably, it is human labour that is transforming the natural world into the built form and the social infrastructure of high-rise housing, the two elements that constitute Singapore's urban landscape.

The invisibility of Singapore's migrant labour is especially striking given how spectacularly visible the product of its labour is. High-rise developments define the city-state while housing is central to the na-tional economy. In Singapore, "real estate not only form[s] a significant portion of national wealth, but also has an important role in the whole economy."[3] The labour of constructing high-rise developments and ser-vicing them every day is not only effaced in the narrative put forward within the HDB Gallery – it is increasingly made invisible in the social landscape of urban space. Aggressive government initiatives mixed with locals' growing intolerance of migrant workers' lives is showcased by pernicious practices that increasingly contain migrant workers to their workplaces and to particular urban spaces.

To further understand this strange intersection of housing, digital technologies, labour, and space, we investigate how the digital technolo-gies that serve to promote the HDB are inserted into labour relations in the housing sector. Digital technologies help employers and government keep migrant workers in line and out of sight, but they are also an important tool for migrant workers to navigate the oppressive geographies of their work and lives. On the one hand, digital technologies facilitate the spatial fix of this regime of migrant labour at the transnational, urban, and corporeal scales, but they also allow migrants to challenge and subvert in small but meaningful ways the conditions and geographies of their work. Information and communications technology (ICT) has become a key infrastructure to deepen unequal power relations and forms of public and private accumulation, but in the hands of migrant labourers, it also enables contestation and compromise.

HOMEWORK

Singapore is being built, quite literally, on migrant labour. More than 40 percent of the city's labour force consists of noncitizens, and the vast majority of this group – upward of 75 percent – are low-skilled, poorly paid, and concentrated in employment sectors considered dirty, dangerous, and demanding. Temporary migrant workers do the vast majority of the work required to build and service Singapore's famed public-housing system. According to official statistics, in 2015 there were 326,000 documented migrant workers in the construction industry and 231,500 documented migrants in domestic work.[4] Together, these two groups make up the majority of work-permit holders in Singapore, a fact that highlights the critical nature of housing, not only to national sentiment but also to the nation's political economy. Yet these same workers are actually barred from accessing the very same housing they help build and maintain. In 2014, the HDB solidified this exclusion by implementing a strict quota for renting or subletting HDB flats to noncitizens.[5]

Migrant workers in the construction sector are drawn mainly from Malaysia, China, Bangladesh, India, Thailand, the Philippines, and Myanmar. Live-in foreign domestic workers (or "maids") come mainly from Indonesia, the Philippines, and Sri Lanka and are employed in 20 percent of Singaporean households. These workers are recruited, paid, housed, and managed differently according to their nationality, with some faring better than others. Building a "home" in Singapore rests on the labour of people separated from their own homes, away from any

▲ **HDB flats.** Photograph by Elizabeth Sibilia.

▶ **Domestic workers meet to discuss issues at TWC2.** Photograph by Samuel Zhengbang. Courtesy of the National Film Board of Canada.

access to familiarity, stability, or comfort. Construction workers are charged high fees for the right to access work; they are paid poorly or unevenly, depending on their nationality; they are frequently injured on the job; and they are treated with contempt and hostility by Singaporean citizens. Workers sometimes return home with debts or debilitating injuries, sometimes both.

Domestic workers also experience poor conditions of work, including widespread physical and sexual abuse, withheld wages, and extensive surveillance.[6] Scholars Brenda Yeoh, Shirlena Huang, and Joaquin Gonzalez argue that "state policy is conceived to ensure that [low-skilled foreign workers] are no more than a transient workforce" that is "subject to repatriation during periods of economic downturn."[7] These authors note that as many as two hundred thousand foreign workers were deported during the 1984–85 recession.

Transient Workers Count 2 (TWC2) is an organization that promotes equitable treatment for migrant workers in Singapore. Shelley Thio, one of its executive committee members, said that she has "come across instances where domestic workers end up owing $4,500 to their agencies ... the average debts women accumulate are between $2,500 and $3,000."[8] Scholars suggest that migrant domestic workers are treated as disposable, as "mobile subjects whose migration and presence in the nation-state are predicated upon a 'use and discard' philosophy designed to ensure that they remain a temporary labour force."[9]

Migrant work programs have become popular with governments all over the world, though Singapore is one of the leaders in the field. Temporary migrant worker programs are highly lucrative for this financialized city-state, because foreign workers on permits have fewer rights than workers with citizenship status. Migrant workers are paid a fraction of what Singaporean citizens earn and are excluded from any form of minimum-wage policy in Singapore. According to the Ministry of Manpower (MOM), "As a matter of national policy, MOM does not prescribe minimum wages for all workers in Singapore, whether local or foreign. Whether wages should increase or decrease is best determined by market demand and supply for labour, skills, capabilities and competency to perform the task."[10]

Though minimum wage is not a national policy, TWC2 reported in 2013 that, on average, Indian and Bangladeshi construction workers were earning as little as SG$700 per month while Chinese construction workers were earning around SG$1,200. These numbers do not include the overhead the labourers owe to the agencies that recruit them, the price of the courses they need to take, and, at times, their housing.[11] That same year, the average monthly income of a Singaporean citizen was SG$3,705.[12] Female domestic workers, who earn as little as SG$200 a month, are also required to pay three months' salary to recruitment agencies to get their contracts.[13]

Alongside these economic advantages, guest worker programs allow for more coercive forms of control over worker mobility, putting a valve on the labour market to help mediate economic swings and their impacts on the domestic economy. Migrant workers enter Singapore through a work-permit system that requires them to migrate without family or dependants, bars them from becoming pregnant or giving birth while in Singapore, and requires approval from the Ministry of Manpower before marriage to a Singaporean citizen or permanent resident.[14]

These programs are also used to manage domestic social and political matters, as the rising numbers of migrant domestic workers here and elsewhere suggests. The city-state issued the earliest permits for domestic workers in the 1970s as a means of facilitating the entry of women with citizenship status into the labour force.[15] This history highlights the complex intersection of gender, class, and nationality by revealing how the domestic rights of women in the labour force were built on the backs of foreign domestic workers.

MIGRANT LABOUR AS SPATIAL FIX

Foreign-worker programs make use of space to solve economic and social problems. Scholars such as Sam Scott use the phrase "spatial fix" to explain how states are expanding their domestic labour markets to include workers from other national contexts who are, in turn, subject to weaker protections and lower wages than citizens of the host state.[16] Conceptualizing migrant labour programs as spatial fix highlights the advantages they bring not only to individual employers but also to entire national economies. The concept was developed by David Harvey and has become a key phrase in human geography, signalling the attempt by private capital or the state to reconfigure space to avert crisis tendencies within capitalism.[17] More specifically, Harvey suggests that declining rates of profit and the tendency towards overaccumulation, which is endemic to capitalist economies, are often managed by making use of particular spatial strategies. Typically, as Daniel Greene and Daniel Joseph explain, "expansion into new or under-exploited geographies becomes a way to dispose of accumulated capital or to create fresh opportunities for new accumulation at faster rates than before."[18] The notion of a spatial fix has been used to theorize a wide array of geographic events, from postwar suburbanization to colonial occupation.

Scott suggests that migrant work programs can usefully be conceptualized as a spatial fix precisely because of the way they mobilize space instrumentally – through the actual physical migration of workers and legal geographies that allow for the partial incorporation of workers without full rights or anything like fair or equal wages. Migrant work programs provide a handy valve for states to control domestic labour markets without having to take into account the needs of migrants. It is precisely this instrumental use of their labour that makes the workers so useful. Foreign migrants, Scott writes, "appear more willing than home-grown workers, or are more compelled by economic necessity and/or their immigration status, to accept such intensification."[19] By virtue of temporarily importing workers from abroad, "employers are better able to control and manage existing workers by using migration to generate highly visible pools of surplus labour."[20]

A spatial fix typically requires the involvement of the state to orchestrate or sanction the fix, and, indeed, migrant work programs rely on the state to craft immigration and labour laws to control the mobility of workers so that they can be made disposable. In Singapore, we might even say that one form of spatial fix – the regime of migrant labour – is being used to fuel another: high-rise urban development.

"FIXING" URBAN SPACE

Migrant work programs require complex practices of spatial control, not only at the international scale of labour recruitment and "repatriation" but also across the multiple spaces in which migrant workers live and work. The urban scale has become particularly acute for struggles over migrant workers' rights. Xenophobia and racism directed against migrant workers have intensified as gentrification in inner-city spaces has led to direct exposure and conflict with the lives of migrant workers. In his research, Daniel Goh emphasizes how "transnational and transient migrants' flows to aspirant global cities, such as Dubai, Hong Kong, Kuala Lumpur, Qatar and Singapore, pose special problems and raise peculiar issues. These are cities ruled by national regimes with an autocratic grip on municipal space."[21]

▲ **Little India.** Photograph by Elizabeth Sibilia.

▼ **Migrant construction workers' shared flat.** Photograph by Elizabeth Sibilia.

It is not only in the HDB Gallery but also in the streets of Singapore that the state is working to make the labour that builds its housing invisible. About four kilometres to the south of the gallery, Little India bustles. The enclave has been home to many foreign workers since Singapore's regime of migrant work programs was initially consolidated. There, migrant workers from India and Bangladesh can find housing in small, private rental dorms above street level. As Goh describes it, "the neighbourhood comprises a tight grid of narrow streets branching off from the main trunk, Serangoon Road, and is made up of densely arranged low-rise shophouses of early twentieth-century heritage architecture."[22] Migrant housing consists of small-scale, privately run, densely populated dormitories over ground-level storefronts.

Even those who don't live in the area gather there on weekends to connect with other migrants and to access the shops, social life, and infrastructure of the area. South Indian and Bangladeshi workers gather in such numbers that they leave "very little room for vehicular and sometimes even pedestrian traffic" as they "meet friends, eat hometown foods and consume services such as repatriating money home, calling loved ones, getting a haircut, and paying for sex."[23] But in recent years, the government has been working to move migrant workers out of the area and to control their conduct and movement if they return.

The practice of housing workers far on the periphery became the norm after the Little India riots in 2013. Before the riots, migrant workers would gather in the central square at Buffalo Road every Sunday for their day off to play cricket and hang out. On December 8, 2013, a foreign worker was killed under the wheels of a private bus company, prompting more than three hundred people to gather and resulting in injuries or damages to fifty-four officers, eight civilians, and a number of emergency vehicles. Given that it was the first riot in Singapore in forty years, responses to it varied. For many, the riot resulted from and symbolized growing frustration over the treatment of migrant workers in Singapore in general. For others, including Debbie Fordyce of TWC2, the riot had been sparked by frustrations about catching the bus on time, a reality for workers whose jobs and living quarters are so inconveniently far apart.[24] The state championed a narrative of the riot as an isolated event, disconnected from growing frustrations among migrant workers, but this narrative has been challenged by community organizers who

▲ **Square in Little India where riots broke out in 2013.** Photograph by Elizabeth Sibilia.

▶ **List of rules put up in the square after the riots.** Photograph by Elizabeth Sibilia.

work in the sector. The riots occurred only one year after Singapore's first strike in twenty-six years – a strike by bus drivers of Chinese nationality who worked for a public transport operator. In a city with approximately 1 million foreign workers, these events, when taken together, were worrying for residents and the Singaporean state.

The riots were thus the occasion for a substantial shift in the urban lives of migrant workers throughout Singapore. Even before the riots, Little India had begun to gentrify, and the city-state welcomed a different class of people to the neighbourhood. The riots offered a pretext for the government to take more aggressive action. It introduced tight restrictions and curfews, and the use of CCTV cameras and security guards rose dramatically. Open spaces where migrant workers once gathered – such as the bus terminal located on the park space, where the riot first took shape – were fractured by new infrastructure. To try to dissolve any form of group congregation, companies that housed migrant workers in dorms outside Little India began to stagger their bus pickup and drop-off times for workers into Little India.

The state has also devised a plan to move their housing out of the downtown altogether. In November 2014, Jolovan Wham from HOME (Humanitarian Organization for Migration Economics) explained to us that the riots led to the formation of a new bill – the Foreign Employee Dormitories Bill – which aimed to create housing for migrant workers so they would stop spending time in public spaces such as Little India.[25] The solution to the inconvenience of migrant lives was the construction of massive, highly securitized purpose-built worker dormitories, thirty kilometres away from the city centre. As Goh highlights, it was only a month after the committee published its report on the Little India riots that the government announced a string of new dormitory projects to house nearly one hundred thousand migrant workers, including a number of megadormitories (of 12,800 to 25,000 beds) built in isolation from the domestic population.[26]

While migrant housing in Little India is overcrowded and overpriced, often infested with cockroaches and bedbugs, and sorely lacking in privacy, many workers insist that the freedom to come and go and the proximity to shops and social life are worth it. The dormitories, by contrast, are equipped with biometrics and CCTV that track and trace the movements of migrants into, out of, and within the facilities. A new housing facility called Tuas View Dormitory, for example, is big enough

CCTV installed at Little India. Photograph by Elizabeth Sibilia.

to house 16,800 men. The complex includes a shopping centre, gym, basketball court, laundry service, medical clinic, dentist, beer garden, and 250-seat cinema. The dorm is also equipped with almost 250 CCTV cameras, and workers must provide their fingerprints to enter and exit the complex. The information goes directly to the police if a crime occurs or for any other reason that security see fit. Closed-circuit television cameras survey not only the entrances but also common areas, including the toilets.[27] This complex was built far from the city centre and is composed of twenty, four-storey blocks. Heavily securitized, the dormitories resemble detention facilities, leading Debbie Fordyce of TWC2 to describe them as fostering "apartheid" conditions (BBC News 2015).[28] The thinking behind them was that if workers had everything they needed in one facility, they would be less likely to venture out to other neighbourhoods or hang around on the streets. The new housing would thus segregate migrant workers "almost completely from the city and the citizens."[29]

DIGITAL DISCIPLINE

Migrant domestic workers who make everyday high-rise life possible for so many Singaporean citizens – through their cleaning, cooking, and child care – also face deeply degraded work conditions. Unlike migrant labourers working in the construction sector, migrant domestic workers can't be contained at the periphery by the city-state because they must live with their employers and work for them around the clock. The requirement that maids "live in" can easily lead to emotional, sexual, and physical abuse, according to Shelley Thio of Transient Workers Count Too.[30] Indeed, their freedom and mobility is aggressively contained but at a different scale than migrant construction workers; they are often hidden in the home at the same time as they are being watched. Digital technologies such as CCTV and smart phones are intensifying disciplinary measures between employers and workers – opening up the possibility of full-time and real-time monitoring of workers' movements and activities and changing the microgeographies of their workplaces.

Singapore is one of the most technologically advanced cities in the world.[31] In 2005, the city-state adopted the Intelligent Nation 2015 (iN2015) masterplan, intended to transform it into "an Intelligent Nation, a Global City, Powered by Infocomm."[32] The program sought to "establish an ultra-high speed, pervasive, intelligent and trusted infocomm infrastructure"; to produce a competitive infocomm industry globally; and to achieve 90 percent home broadband usage and 100 percent computer ownership in homes with school-aged children. In 2015, household broadband penetration was at 102.6 percent and mobile phone penetration at 148.4 percent – each ICT statistic exceeding that of Singapore's population.

Despite widespread ICT use among Singaporean citizens, migrant workers until recently had only limited access to mobile phones and other ICT because of their high costs. In recent years, however, foreign domestic workers have become target customers for communications companies as low-cost phones and plans proliferate.[33] Yet access to digital technologies is not a simple story that sees workers having more connection to loved ones or autonomy in their everyday lives; it is also one of extended social control and worker discipline. The ubiquitous use of ICT exemplifies some of the more repressive aspects of these technologies; it has contributed to the blurring of work and nonwork time by allowing employers to contact their employees at all times of the day and night.

In focus groups, Filipina and Indonesian domestic workers told us repeatedly about how digital technologies allowed their employers to invade their small bits of nonwork time. The women recounted stories of employers using cellphones, closed-circuit television, and other ICT to watch them from a distance. This surveillance took place within a context of already deeply circumscribed privacy. Domestic workers are required to live on site with their employers, and HOME reports that it has "seen many cases where workers are not provided with a private room and sleep in a common living area, along a corridor, outside on a balcony or sharing a bedroom."[34]

Many Singapore-based employment agencies are now equipping foreign domestic workers with SIM (subscriber identity module) cards before they are deployed in order to cater to employers' demands and facilitate control over employees.[35] In our conversations with domestic workers, we heard story after story of employers giving cellphones to domestic workers so that they could reach them at all times. Some employers took the phones away during work hours to limit domestic workers' social time and connectivity. Other workers spoke with their employers regularly through Facebook Messenger, a website that workers also use to keep in touch with their family and friends back home. For workers who could keep a phone with them throughout the day, knowing their use pattern was being monitored (Facebook makes it possible to see when people are online), was enough to keep them from using the phone for anything more than work.

ICT has indeed invaded the domestic workplace. HOME reports that "some employers also install surveillance cameras within the household, including in the toilet and sleeping areas, to monitor the movements and behaviour of the domestic worker."[36] One domestic worker told us her employer installed surveillance cameras in the kitchen of the apartment and, when away, locked her in the kitchen so her activities could be monitored from afar. The domestic worker described the insidious feeling of being constantly watched and how her employer would phone her almost immediately if she stepped out of the camera's view to ask where she was and what she was doing.

The most surprising stories we heard were also the most dehumanizing. For a number of the workers, Skyping with employers who were out of town had become a regular part of their jobs. But when their employers called, they asked to speak to the family pet. One worker described the painful experience of being asked to move out of view so the employer's dog could be more visible.

Tensions about surveillance have only become more acute since employers now have to pay a steep levy of SG$5,000 to the Ministry of Manpower, which is forfeited if the worker breaches the conditions of her work permit.[37] Given the financial overhead, and the often tenuous relationship between employer and employee, some employers have been aggressive in their surveillance of workers for fear of losing their security bond.[38] Breaches include leaving the employer, pregnancy, or marriage to a Singaporean or permanent resident.[39] The fact that pregnancy or intimate relations with Singaporean citizens is considered a breach means that domestic workers' personal lives are subject to often intensive surveillance by their employers. This intrusion is in addition to invasive procedures imposed by the government – for instance, each worker has to take a pregnancy test upon application to work in Singapore and must undergo regular pregnancy tests to ensure they don't give birth to children in the country.

With the click of a button on the MOM website, employers can also make a sudden and unilateral decision to cancel the employment of a domestic worker. This action places the worker on a blacklist with no form of redress. As Shelley Thio, executive committee member of TWC2, explains, the worker is given no trial or specific reason for the termination of her contract, and once she has been blacklisted, she can't leave her next employer, no matter how bad her situation becomes.[40]

(DIGITAL) SURVIVAL AND THE SPATIAL FIX

Despite the intensification of digital surveillance in the city and at worksites, access to ICT can also facilitate a greater sense of connectivity and autonomy for workers. Digital technologies are helping workers navigate and at times even reconfigure their relationship to coercive geographies. They allow migrant workers to access spaces that they can't access in other ways, and they can provide tools for mediating the unequal power relations they experience in their workplaces. For migrant domestic and construction workers, access to ICT is coveted precisely because it allows them to navigate the complex geographies of their everyday lives. In our conversations with them, some described these technologies as precious because they allowed them to connect with their families back home over Skype, to send remittances more easily and at lower cost, and to connect with other migrants locally. For migrant workers, ICT has arguably become increasingly critical for their survival across the complex and often alienating geographies of their everyday lives.

Scholars have noted that the pervasiveness of ICTs has, in part, bridged the temporal and spatial gap caused by migration.[41] In the case of domestic workers, mobile phones have enabled them to fulfill their parenting and care duties despite their physical absence.[42] ICTs play a key role in facilitating the upkeep of a broad range of family ties. They have brought about new ways for families to live together transnationally.[43] Through technological media such as Skype and Facebook, transnational migrant families can transmit family memories, express their identities, and establish a sense of connectedness.[44] In her exploration of long-distance intimacy, Deirdre McKay argues that ICTs can indeed constitute a valid form of intimacy.[45]

But ICTs are also extending the work day in complicated ways for many migrant workers. Having access to cellphones can increase the pressure on women to engage in their children's lives back home more readily. This is a phenomenon Rhacel Salazar Parreñas has labelled "transnational mothering" and is evident in the research of Platt and colleagues. Their respondents recounted keeping in close communication with their families back home to oversee their children's activities such as making sure "they're doing their homework."[46] These duties are on top of an average fourteen-hour work day.[47]

The sense that ICT can complicate ties to family back home was echoed in the focus groups we held with construction workers who had been injured on the job site and were awaiting some kind of medical treatment or legal process before they could return to work or be repatriated. One worker was anxious that his family would discover his injured status by finding him online during work hours. ICT first and foremost allows for the surveillance of workers by employers, but it also allows their own families to monitor them, thus instigating more stress or emotional labour. Yet, as Symon James-Wilson argues in the final essay of this book, ICT use shouldn't solidify narratives of repression or of redemption; it's important to focus instead on the varying technologies (old and new) that workers use to survive or subvert the system.

In some cases, ICT allows workers to counter the surveillance imposed on them by these very same technologies. For instance, we met with a group of Chinese construction workers who were in limbo in Singapore following injuries they sustained while at work. One man spoke about how he had actively used his cellphone to record conversations with his employer. He also kept notes on his cellphone about events in his workplace in case he needed the information in his case with the Ministry of Manpower. His actions mark creative "sous-veillance" through ICT while also registering the troubling state of labour relations in Singapore, especially disputes over workplace safety.

Through seemingly ordinary uses of ICT, foreign domestic workers have the ability to regain "subjectivity, connection, and freedom," and they often do so in a manner that subverts restrictions placed on them by their employers.[48] One example is the use of online-dating sites and apps, which one of the women spoke about with enthusiasm. Though migrant workers are banned from marrying and having children with Singaporean citizens, many women who do not have families or partners back home use online-dating sites to meet men elsewhere in the world. Some of these relationships then stay online, as many don't have enough time off to take them off-line. Others spoke about travelling to meet men in different parts of the world and of carrying on these relationships for many years.

Digital technologies are also facilitating formal and informal kinds of worker organizing. In Singapore, ICT has been at the centre of outreach, education, and organizing efforts with migrant workers. For instance, Indonesian and Filipino family networks, informal organizations formed by domestic workers, are a way to come together around shared experiences in Singapore and to create a local sense of community. The networks help women get settled into life in Singapore in the realm of domestic labour; they also create longer-lasting community networks for women missing family and friends back home. The networks offer training on how to use the internet and Microsoft Word, and they also run workshops on finance, photography, languages, and flower arranging, and they teach workers how to sell some of the items they make online. These networks focus on teaching workers how to communicate with their families back home via Skype, Yahoo, Facebook, WhatsApp, and so on, because most workers only know how to use a computer to use Facebook.

One of the more intriguing uses of these technologies that we learned about from migrant workers was the way they were facilitating

knowledge sharing and even action in response to Singapore's oppressive labour conditions. They described their use of the internet to communicate their experiences to others back home. A number of the Bangladeshi workers disclosed that they had written to people back home who were planning to come to Singapore to tell them not to come. They warned them of the huge overhead costs they would have to pay to the agencies that brought them over, and they revealed the lies these agencies had told about the kinds of work they would be doing when they got to Singapore. Workers described to friends and relatives back home their experiences of being employed in areas of work they had no training for, despite being promised jobs in their fields, and they connected this misrepresentation to the workplace injuries they and their colleagues were experiencing.

Domestic workers also related stories back home to help prepare friends and family for what to expect when coming to Singapore, and they, too, often explicitly discouraged others from coming. They said their stories had directly contributed to the declining numbers of workers coming from the Philippines and Indonesia and in part explained the Singaporean government's recruitment in Myanmar. Although there are many factors that contributed to this shift, they no doubt had an impact. Estimates by embassies and employment agencies suggest a 50 percent increase in domestic workers from Myanmar working in Singapore

between 2013 and 2015.[49] In their research, Platt and colleagues recount similar experiences. One woman, for instance, said she followed the Hong Kong branch of the Indonesian Migrant Workers Union on Facebook to gain information on domestic workers' rights more broadly.[50]

• • • • •

Migrant workers are the linchpin of Singapore's economy. Migrant workers and their labour are critical to building the physical landscape of the city, which is at the centre of its national economy and identity. Migrant work is a spatial fix in that it is through a particular spatial arrangement that the state and capital aim to solve a political economic problem. By expanding labour recruitment to the transnational scale, the government can simultaneously subsidize individual employers and the national economy while maintaining strict national boundaries regarding who can enjoy full political and economic rights. Migrant labour is a spatial fix, and so, too, is the high-rise landscape that the workers build and maintain. Indeed, the physical construction of the city-state and its financialization are key to Singaporean economic development and perhaps to global accumulation more broadly.[51]

Despite or perhaps because of their importance, migrant workers are rendered invisible in Singapore's streets and in the stories the city-state tells of itself. The HDB Gallery offers romantic accounts of Singaporean families building the nation without any mention of the people actually doing the work. Similarly, official accounts of the history of housing in Singapore proudly recount the emergence of social housing plans in postcolonial Singapore, but again without a mention of the migrants whose hyper-exploited labour made it possible.[52] Finally, scholarly work about the Singaporean state's labour practices treats the topic of migrant work as if it were not a polite topic. As Daniel Goh suggests, "Most local academics avoid tackling foreign worker issues, turning their attention instead to the identity politics of history and heritage and questions of protecting local privilege in an academic arena where they are the minority."[53] There are, of course, exceptions, such as the work of Brenda Yeoh and Maria Platt, who raise pressing questions about the state of work and labour in Singapore.

Efforts to make migrants invisible are also apparent in the material space of the city-state, in state policies to move workers out of the inner city and into self-contained and purpose-built dormitories on the periphery. The invisibilization of migrant workers is part of their marginalization; it makes it increasingly difficult for them to be recognized and compensated for their labour and sacrifices, and it helps to sustain dehumanizing forms of racism. Rendering them invisible also helps perpetuate a national fantasy in which the high-rise buildings upon which the city-state and its economy and identity have been magically built, clean and care for themselves.

The mobility of workers in space is also a site of struggle in Singapore, particularly at the transnational scale, where strict controls on their movements are enacted by governments and employers. But at the local level, the comings and goings of migrant workers in the city and at the work site are also the target of scrutiny and surveillance, whether in dormitories or in the employer's home.

Woven through it all are digital technologies, which facilitate control of workers' mobility at several scales – the transnational, the urban, and the corporeal. Digital technologies are instrumental in the recruitment of migrants: they manage the collection and storage of their citizenship data, and they enable employers to access the programs, the permits, and the workers themselves. They are then mobilized to monitor migrants' mobilities within urban space and the worksite. Yet migrants also rely on them to navigate their own lives and protect and serve their interests across space and at different scales. The digital technologies that structure the daily life of migrant workers in Singapore are used both as a disciplining strategy, and as a means of survival.

NOTES

1 Move through a 3D rendering of the Housing and Development Board (HDB) Gallery in Singapore on their website at https://www.hdb.gov.sg/cs/infoweb/about-us/livingspace.

2 HDB Gallery, "LIVINGSPACE," https://www.hdb.gov.sg/cs/infoweb/about -us/livingspace.

3 Anne Haila, "Real Estate in Global Cities: Singapore and Hong Kong as Property States," *Urban Studies* 37, 12 (2000): 2241.

4 Ministry of Manpower, "Foreign Workforce Numbers," March 18, 2016, http://www.mom.gov.sg/documents-and-publications/foreign-workforce -numbers.

5 As of January 16, 2014, the maximum proportion of flats that can be sublet to non-Malaysian noncitizen subtenants is 8 percent per neighbourhood. The HDB's website states that "the quota aims to prevent the formation of foreigner enclaves in public housing estates."

6 Maria Platt, Kristel Anne Acedera, Grace Baey, Khoo Choon Yen, Brenda Yeoh, and Theodora Lam, "Renegotiating Migration Experiences: Indonesian Domestic Workers in Singapore and Use of Information Communication Technologies," *New Media and Society* 18, 10 (2016): 2207–23; Shirlena Huang and Brenda S.A. Yeoh, "Emotional Labour and Transnational Domestic Work: The Moving Geographies of 'Maid Abuse' in Singapore," *Mobilities* 2, 2 (2007): 195–217; and Amarjit Kaur, "The Global Labour Market: International Labour Migration in Southeast Asia since the 1980s," in *Wage Labour in Southeast Asia since 1840: Globalization, the International Division of Labour and Labour Transformations,* ed. Amarjit Kaur (London: Palgrave Macmillan, 2004), 197–230.

7 Brenda S.A. Yeoh, Shirlena Huang, and Joaquin Gonzalez, "Migrant Female Domestic Workers: Debating the Economic, Social and Political Impacts in Singapore," *International Migration Review* 33, 1 (1999): 117.

8 "Buy a Discount Maid at Singapore's Malls," *Al Jazeera,* June 27, 2014.

9 See Platt et al., "Renegotiating Migration Experiences," 2209. See also Brenda S.A. Yeoh, "Bifurcated Labour: The Unequal Incorporation of Transmigrants in Singapore," *Tijdschrift voor Economische en Sociale Geografie* 97, 1 (2006): 26–37; and Brenda Yeoh and T.C. Chang, "Globalising Singapore: Debating Transnational Flows in the City," *Urban Studies* 38, 7 (2001): 1025–44.

10 Ministry of Manpower, "Is There a Prescribed Minimum Wage for Foreign Workers in Singapore?," http://www.mom.gov.sg/faq/work-permit-for-foreign

-worker/is-there-a-prescribed-minimum-wage-for-foreign-workers-in
-singapore.

11 Transient Workers Count Too (TWC2), "Low Pay May Deter Foreign Workers," news flash, January 6, 2013, http://twc2.org.sg/2013/01/06/low
-pay-may-deter-foreign-workers/.

12 Ministry of Manpower, "Summary Table: Income," June 10, 2016, http://stats.mom.gov.sg/Pages/Income-Summary-Table.aspx.

13 Kaur, "The Global Labour Market."

14 Ministry of Manpower, "Work Permit Conditions," May 5, 2015, http://www.mom.gov.sg/passes-and-permits/work-permit-for-foreign-worker/sector-specific-rules/work-permit-conditions.

15 Yeoh, Huang, and Gonzalez, "Migrant Female Domestic Workers."

16 Sam Scott, "Labour, Migration and the Spatial Fix: Evidence from the UK Food Industry," *Antipode* 45, 5 (2013): 1090–109.

17 See David Harvey, *The Limits to Capital* (Chicago: University of Chicago Press, 1982) and "Globalization and the 'Spatial Fix,'" *Geographische Revue* 2 (2001): 23–30.

18 Daniel Greene and Daniel Joseph, "The Digital Spatial Fix," *tripleC: Communication, Capitalism and Critique – Journal for a Global Sustainable Information Society* 13, 2 (2015): 223–47.

19 Scott, "Labour, Migration and the Spatial Fix," 1095.

20 Ibid.

21 Daniel P.S. Goh, "The Little India Riot and the Spatiality of Migrant Labor in Singapore," *Society and Space* (2014), https://societyandspace.org/2014/09/08/the-little-india-riot-and-the-spatiality-of-migrant-labor-in
-singapore/.

22 Ibid.

23 Ibid.

24 Debbie Fordyce, interview with authors, TWC2, Singapore, November 2014.

25 Jolovan Wham, interview with authors, November 2014. According to the Ministry of Manpower, the bill was passed in January of 2015. https://www.mom.gov.sg/newsroom/announcements/2015/foreign-employee
-dormitories-bill.

26 Goh, "The Little India Riot."

27 HOME and MWRN, "A Submission by Humanitarian Organization for Migration Economics (HOME) and Migrant Worker Rights Network (MWRN) for the 23rd Session of the Universal Periodic Review," November 2015.

28 "Singapore Is Keeping an Eye on Its Migrant Workers," *BBC News*, April 14, 2015, http://www.bbc.com/news/business-32297860.

29 Goh, "The Little India Riot."

30 "Buy a Discount Maid."

31 Platt et al., "Renegotiating Migration Experiences," 2.

32 Ibid.
33 E.C. Thompson, "Mobile Phones, Communities and Social Networks among Foreign Workers in Singapore," *Global Networks* 9, 39 (2009): 359–80; and Platt et al., "Renegotiating Migration Experiences."
34 Solidarity for Migrant Workers, "A Joint Submission by Members of Solidarity for Migrant Workers for the 11th Session of the Universal Periodic Review," May 2011, http://apmigration.ilo.org/resources/a-joint-submission-by-members-of-solidarity-for-migrant-workers-for-the-11th-session-of-the-universal-periodic-review-may-2011.
35 C.W. Aw, "More Employers Unhappy over Maids' Mobile Phone Use," *Straits Times*, July 12, 2015, cited in Platt et al., "Renegotiating Migration Experiences."
36 Solidarity for Migrant Workers, "A Joint Submission."
37 Ministry of Manpower, "Work Permit Conditions," May 5, 2016, http://www.mom.gov.sg/passes-and-permits/work-permit-for-foreign-worker/sector-specific-rules/work-permit-conditions.
38 Hoi-yu Polly Poon, "Right to Privacy and Surveillance in a Technology-Mediated Society," *Cultural Studies* 36 (2013), http://commons.ln.edu.hk/mcsln/vol36/iss1/8/; and J. Seow, "Some Bosses 'Reluctant' to Give Maids Weekly Day Off," *Straits Times*, April 16, 2014.
39 *Employment of Foreign Manpower Act*, c 91A, cited in Platt et al., "Renegotiating Migration Experiences."
40 Shelley Thio, interview with authors, November 2014.
41 J.V.A. Cabanes and K.A.F. Acedera, "Of Mobile Phones and Mother-Fathers: Calls, Text Messages, and Conjugal Power Relations in Mother-Away Filipino Families," *New Media and Society* 14, 6 (2012): 916–30; Mirca Madianou, "Migration and the Accentuated Ambivalence of Motherhood: The Role of ICTs in Filipino Transnational Families," *Global Networks* 12, 3 (2012): 277–95, and "Polymedia Communication and Mediatized Migration: An Ethnographic Approach," in *Mediatization of Communication*, ed. Knut Lundby (Berlin: De Gruyter, 2014), 323–48; Daniel Miller and Mirca Madianou, "Should You Accept a Friends Request from Your Mother? And Other Filipino Dilemmas," *International Review of Social Research* 2, 1 (2012): 9–28; Rhacel Salazar Parreñas, *Children of Global Migration: Transnational Families and Gendered Woes* (Stanford, CA: Stanford University Press, 2005); E.C. Thompson, "Mobile Phones, Communities and Social Networks among Foreign Workers in Singapore," *Global Networks* 9, 39 (2009): 359–80; and Cecilia Uy-Tioco, "Overseas Filipino Workers and Text Messaging: Reinventing Transnational Mothering," *Continuum* 21, 2 (2007): 253–65.
42 A. Fresnoza-Flot, "Migration Status and Transnational Mothering: The Case of Filipino Migrants in France," *Global Networks* 9 (2009): 252–70; Deirdre Mckay, "'Sending Dollars Shows Feeling': Emotions and Economies in Filipino Migration," *Mobilities* 2, 2 (2007): 175–94; Miller and

Madianou, "Should You Accept a Friends Request"; Madianou "Migration and the Accentuated Ambivalence of Motherhood"; Parreñas, *Children of Global Migration* and *Servants of Globalization: Migration and Domestic Work,* 2nd ed. (Palo Alto, CA: Stanford University Press, 2015); and Uy-Tioco, "Overseas Filipino Workers and Text Messaging."

43 Mihaela Nedelcu, "Migrants' New Transnational Habitus: Rethinking Migration through a Cosmopolitan Lens in the Digital Age," *Journal of Ethnic and Migration Studies* 38, 9 (2012): 1339–56.

44 J. Hamel, "Information and Communication Technologies and Migration," United Nations Development Programme, New York, Human Development Research Paper No. 392009; and José Luis Benítez, "Salvadoran Transnational Families: ICT and Communication Practices in the Network Society," *Journal of Ethnic and Migration Studies* 38, 9 (2012): 1439–49.

45 Mckay, "Sending Dollars Shows Feeling."

46 Parreñas, *Children of Global Migration,* and Platt et al., "Renegotiating Migration Experiences," 11.

47 Platt et al., "Renegotiating Migration Experiences."

48 T.T.-C. Lin and S.H.-L. Sun, "Connection as a Form of Resisting Control: Mobile Phone Usage of Foreign Domestic Workers in Singapore," *Media Asia: An Asian Communication Quarterly* 37, 4 (2010): 190; and K. Ueno, "Strategies of Resistance among Filipina and Indonesian Domestic Workers in Singapore," *Asian and Pacific Migration Journal* 18, 4 (2009): 497–517.

49 HOME and MWRN, "A Submission."

50 Platt et al., "Renegotiating Migration Experiences," 9.

51 Anne Haila, "Real Estate in Global Cities: Singapore and Hong Kong as Property States," *Urban Studies* 37, 12 (2000): 2241–56; and Henri Lefebvre, *The Urban Revolution* (Minneapolis: University of Minnesota Press, 2003).

52 Lily Kong, *Conserving the Past, Creating the Future: Urban Heritage in Singapore,* Urban Redevelopment Authority (Singapore: Singapore Press Holdings, 2011).

53 Goh, "The Little India Riot."

Skyline of Dreams

Grace Baey

IF YOU VISUALIZE THE modern Singapore skyline, the Marina Bay Sands Integrated Resort, with its iconic spaceship-like deck and infinity pool, likely comes to mind. This is the place that forty-five-year-old Mustafa, a migrant construction worker from Bangladesh, has been wanting to visit for some time now. "It costs twenty-three dollars just to get up there – too much money!" he said. "You know what's funny? When we're building these buildings, we get free access to roam about anywhere. But when it's completed, they put a security guard outside, and we can't go in."

Mustafa's statement offers a poignant description of how migrant construction workers are often treated as an invisible class of labourers – or what Anoma Pieris terms hidden hands – within the city.[1] The Singapore skyline was meant to reflect the dreams of its people. As Prime Minister Lee Hsien Loong articulated in 2005, "It will be a city in our image, a sparkling jewel, [and] a home for all of us to be proud of."[2] But beneath this landscape lies a myriad of hidden dreams, along with the hidden costs that migrant workers pay to build the city.

BIG DREAMS, HEAVY INVESTMENTS

When thirty-eight-year-old Sumon contemplated the thought of working in Singapore, his wife encouraged him. She said, "Go there. Many people have been there. Your friend has changed his life, and now it is your turn to change our fate. You will come back, and you will have some money." He was helping out with his father's business in Bangladesh, but it did not provide "a perfect earning source to change [his] lifestyle."[3]

He had bigger dreams of buying land, improving the family house, and starting his own business.

Migrating to Singapore for work was, of course, not without its costs. In 2015, Bangladeshi construction workers paid an average of BDT$393,275 (US$4,987) in predeparture fees, either to an agent or training centre, to access mandatory skills training, testing, and job-placement services in Bangladesh. To finance these fees, workers borrowed on average BDT$258,423 (US$3,278), or 65.7 percent of their total placement fee from a range of sources, such as relatives or the bank.[4]

While debt offered a viable strategy to finance one's migration journey, the sheer amount also meant that the majority of workers' initial earnings had to be siphoned towards debt repayment. In this context, ensuring a steady income flow is especially crucial, and Bangladeshi migrants take, on average, sixteen and a half months to fully repay their initial loans.[5] These investments entail a significant gamble, since work permits in the construction sector are typically issued on a one-year basis with no guarantee of renewal. As much as these investments are often deemed highly risky, workers tend to justify them by weighing the costs against the longer-term benefits of receiving larger salaries from overseas work in Singapore.

In Sumon's case, he was uncertain that his university degree would secure him a viable job in Bangladesh. "If I started a job matching my qualifications, I would get a maximum of BDT 10,000 ($126 USD) per month, which is a very poor price for [my] livelihood."[6] He made plans to work in Singapore for ten years in the hopes of saving money to purchase land, renovate his family home, and start a business. At present, he earns a basic wage of twelve US dollars per day, and he clocks approximately thirty hours of overtime weekly. This gruelling schedule has enabled him to remit a monthly average of US$550 back home.

PRECARIOUS WORK, UNFORESEEN COSTS

While the construction industry provides a significant source of employment for many low-waged migrants, its precarious nature as dirty, dangerous, and difficult work is typically accentuated by flexible labour-market policies, high recruitment and job-placement fees, and restrictive visa regimes that tend to limit migrants' access to labour mobility and social protection.

In Singapore, workers' visas are tied to a single employer, and workers are not permitted to seek a change of employer (with the exception of the construction sector, where prior written consent by one's current employer is needed). Since workers often take on significant loans to secure overseas job opportunities, many sometimes choose to endure harsh and unsafe employment conditions rather than risk the possibility of early repatriation. The industry remains the top contributor to workplace fatalities in Singapore, accounting for 57 percent of all fatalities reported in 2013.[7]

Mizan, age twenty-three, broke both wrists when he fell from scaffolding while building a flyover. Before the accident, he was clocking an average of fourteen hours daily. He believes that fatigue stemming from excessive overtime work and poor workplace-safety enforcement were the main contributing factors to the incident. As a new employee in the company, he felt he had little bargaining power to negotiate working terms with his supervisor, for fear that his work permit might be cancelled prematurely and that he would be sent home in debt.

Although workers are covered under Singapore's Work Injury Compensation Act, the claims process can be arduous and lengthy, ranging from three or four months to over two years, depending on the nature of the injury and the complexity of the case. At times, employers have reportedly tried to conceal workplace accidents to evade penalty measures by the government that restrict hiring practices for those with poor workplace-safety records. In Mizan's case, his employer tried to coax him to provide false details of the accident, alleging that he had fallen behind a lorry instead of from the scaffolding.

Faced with limited employment options stemming from injury, workers often place an enormous value on the compensation they hope to receive. As thirty-one-year-old Mostafa said, "I will build my whole life on that money [because] a huge damage has been done to my physical body. If [the sum] is large, then I can do something after returning to my country."[8]

AMBIVALENT FUTURES

Insofar as low-waged migrant workers provide a necessary source of cheap and flexible labour for Singapore to realize its goal of being a thriving global city, most of the costs that Bangladeshi migrants pay to

Akram Hossain looks at a picture of his wife as he anticipates the opportunity to return home after a lift-repair accident left him with two severed fingers. Photograph by Grace Baey.

build this dream are typically hidden. For some, these costs are never fully recovered, especially when workers suffer permanent injuries stemming from workplace accidents during tight project deadlines in the construction industry. In these cases, migrant dreams of securing better livelihoods have to be put on hold or are never realized.

Arguably, Singapore's skyline is being built on a multitude of dreams, both told and untold. Is it a place where different dreams can thrive? In Sumon's view, migrant workers are merely treated as labouring bodies within the city: "I have value till I can work. But when I won't be able to work, I will have no value in Singapore."[9] Regardless, many Bangladeshi migrants continue to take the gamble to invest in migration for construction work as a livelihood strategy to invest for a better future – helping to build a city's dream, even as they strive for their own.

NOTES

The interview material from this article was drawn from a research study titled "Migration and Precarious Work: Negotiating Debt, Employment, and Livelihood Strategies amongst Bangladeshi Migrant Men Working in

Singapore's Construction Industry," by Grace Baey and Brenda S.A. Yeoh, under the Migrating Out of Poverty Research Programme Consortium, funded by UK Aid.

1 Anoma Pieris, *Hidden Hands and Divided Landscapes: A Penal History of Singapore's Plural Society* (Honolulu: University of Hawai'i Press, 2009).

2 Lee Hsien Loong, National Day Rally speech, NUS University Cultural Centre, August 21, 2005, http://www.nas.gov.sg/archivesonline/speeches/view-html?filename=2005082102.htm.

3 Grace Baey and Brenda S.A. Yeoh, "Migration and Precarious Work: Negotiating Debt, Employment, and Livelihood Strategies amongst Bangladeshi Migrant Men Working in Singapore's Construction Industry," Migrating Out of Poverty Working Paper No. 26, University of Sussex, Brighton, 2015, 16.

4 Ibid., 16.

5 Ibid., 24.

6 Ibid., 16.

7 WSH Institute, *Workplace Safety and Health Report: National Statistics, January to June 2014*, 2014, https://www.wshc.sg/files/wshc/upload/cms/file/2014/Workplace_Safety_and_Health_Report_January_to_June_2014.pdf.

8 Grace Baey and Brenda S.A. Yeoh, "Migration and Precarious Work: Negotiating Debt, Employment, and Livelihood Strategies amongst Bangladeshi Migrant Men Working in Singapore's Construction Industry," 29.

9 Ibid., 32.

Sunny Island Set in the Sea

Charmaine Chua

THE GROUND ON WHICH I stand did not exist fifteen years ago. I am standing on the forty-seventh floor of a glass building near Singapore's Central Business District, on a skyscraper whose foundation was once ocean. Below me is artificial ground solid enough to hold the weight of an endless profusion of high-rise buildings. Beyond the glass windows, I gaze towards the coast at a large oblong piece of land that protrudes three kilometres into the ocean. The friend whose offices I am visiting tells me that it is Phase 3 in the massive US$3.5 billion Pasir Panjang port development project. In the distance, dredging ships pull sand from the seabed. Barges dump load after load of sand into the water. Vast tracts of pulverized, dredged, and piled silt sit in heaps on a perfectly rectangular coastline. Land, in other words, is being created before my very eyes.

"Because the port thrives, so Singapore thrives," Prime Minister Lee Hsien Loong declared at the unveiling of the terminal three months later, articulating a common refrain in the national imaginary: if the economic and social future of this tiny nation-state hinges on the continuous expansion of its markets and working population, then it will require the expansion of the spaces in which they operate.[1]

Until recently, territory has been regarded largely as an unassailable material limit of sovereignty. In the modern conception, to rule is to have authority over a bounded space, to exercise control over a people within the seemingly permanent features of landed territory. Of course, rulers have long sought to expand the zones of operation through which they could exercise their power, shifting the borders and boundaries of rule through acts of bloody conquest and dispossession. But what if, today, rather than taking over already-existing territory one could literally create it? What if rather than being an immovable geological fact, land is actually a mobile commodity conjured through an accretion of the most granular of forms – sand?

As a desperately land-scarce nation, the island state of Singapore has, for much of its short history, been engaged in what is known as land reclamation projects in order to increase the living and working space of the island. In the fifty years since its independence, its population has more than doubled, requiring the continuous construction of both private condominiums and the high-rise public housing that serves 80 percent of the population. Vertical growth, however, has not been enough to sustain a burgeoning populace, so the state has sought horizontal expansion since its independence. Singapore's land area has grown from 581.5 square kilometres in the 1960s to 723.2 square kilometres today, an increase in territory of almost 24 percent. By 2033, the government plans to increase its land area by another 100 square kilometres, making the island a full 30 percent larger than its original size. Most of Singapore's reclaimed land occupies patches of sea that were once part of the Singapore Straits that separate the island from Indonesia, demanding a shift of maritime boundaries every time new territory is claimed.

To achieve these monumental acts of creation, colossal amounts of sand are required. Sand may seem to be a fairly innocuous particulate in its granular form – experienced, for instance, as the grains that stick under your swimsuit after a day at the beach or as the shifting shore under your feet. But this granularity is precisely what makes sand a valuable medium. Both fluid and solid, sand possesses a softness and scalability that allows for its easy transportation across great distances, whether by truckload, bargeload, or spadeful. In each of these applications, sand, itself a form of territory, skews and transforms notions of territorial space, conquering vertical space with concrete or aiding the annihilation of distance with technology. Yet none of these applications reconfigure territoriality more than when sand is terraformed in enough quantity to turn it into the most foundational infrastructural form: land.

Acquiring enough sand to create these new landmasses is a colossal task. To supply itself with reclamation material, Singapore levelled most of its hills in the 1960s, transforming an undulating island into a largely flat surface. Then, it dredged its coastal seabed. But local resources were barely sufficient to support the city-state's massive need, and so Singapore

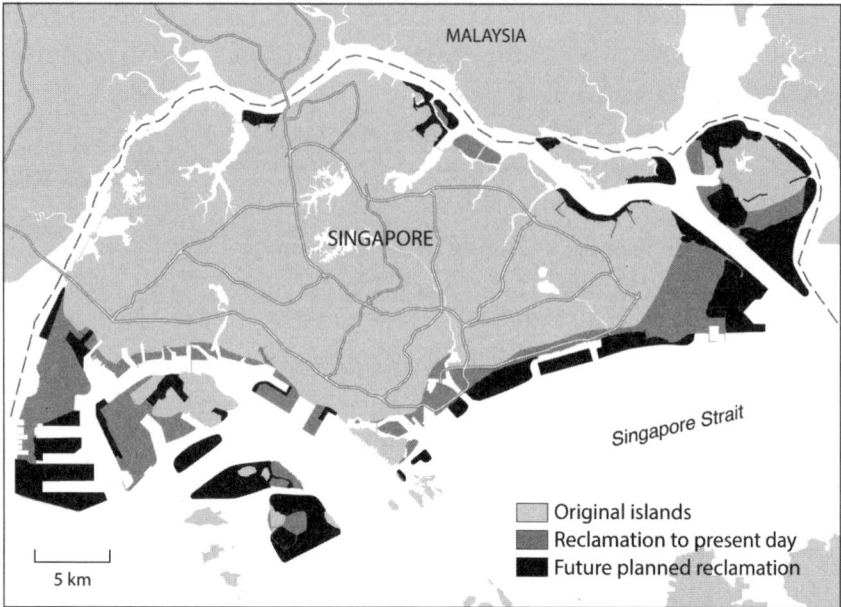

Adapted from map found at Blogs@NTU, "Impact of Land Reclamation in Singapore," https://blogs.ntu.edu.sg/hp331-2014-10/?page_id=7.

began importing sand from neighbouring countries starting in the 1970s. In the last twenty years alone, it has imported a reported 517 million tonnes of sand, making it by far the largest importer of sand worldwide.[2] To give this mammoth figure some context, terraforming 1.5 square kilometres of new ground requires 37.5 million cubic metres of sand fill. This is the equivalent of 1.4 million dump trucks' worth of sand – a line of trucks so long that it would snake from New York City to Los Angeles and back again.

Until 2007, the largest sources of sand imports were Indonesia, Malaysia, and Vietnam, but the depletion of marine life, landslides, and river erosion that followed sand mining and the erasure of at least twenty-four Indonesian islands since 2005 prompted these countries to restrict or ban exports of sand to Singapore.[3] Today, most of Singapore's sand fill needs are supplied by Myanmar and Cambodia, which have, in turn, begun to report the devastating effects of the sand trade on local populations.[4] Accompanying Singapore's increased demand for sand has been a huge boom in illegal sand mining. In 2012, a total of 120 million tonnes of sand were reported missing, a variance in import-export figures that suggests that illegal mining and black-market smuggling are playing an integral role in the expansion of a city-state that often prides itself on being clean of corruption.[5]

Accounts of sand smuggling have accordingly become preoccupied with two aspects of the sand trade. First, noting the massive amounts of sand changing hands through illegal means, journalists such as Chris Milton expose its seedy underbelly, with the implicit hope that an awareness of the sand trade's illicit economies will bolster efforts at better regulation, placing limits on the state's capacity to destroy crucial environmental resources.[6] Second, scholars note that large-scale movements of sand throw into question the legality of practices of terraforming under international law, as they exacerbate geopolitical conflict by encroaching on the territorial jurisdiction of neighbouring countries. These geopolitical tensions are most evident in China's terraforming excursions in the South China Sea, a form of "reclamation" that has provoked heated territorial water disputes between multiple nations and threats of retaliation and even war. Both approaches attempt to reign in the expansionist tendencies of state terraforming by appealing to accountability under international law. By doing so, they misrecognize the issue as a jurisprudential one rather than one in which geopolitical imperatives are shaping and reconfiguring the law itself.

Take, for example, Singapore's neighbour across the Johor Strait. For decades, since sand mining began, Malaysia has bristled at the slow creep of Singapore's soil into its coastal waters, demanding at each instance a reapportioning of the lucrative territorial waters and sea trade lanes that lie between the two states. Yet these demands have relatively little legal backing. Under the United Nations Convention on the Law of the Sea, Singapore can legally "reclaim" sovereignty around existing islands, reefs, and archipelagos. What the ostensible legitimacy of land reclamation practices suggests is that we are experiencing a form of volatile sovereignty quite different from pre-existing modes. With land reclamation, states are geophysically engineering the globe at a scale that shifts the very ground on which sovereignty is thinkable. The geophysical follows the geopolitical rather than the other way around. Joshua Comaroff notes, for instance, that because the "physical basis of the state can be incrementally eroded or expanded," land reclamation inaugurates a "flow of territory" quite distinct from other forms of territorial expansion such as war, military occupation, or colonial expansion.[7] What the unchecked phenomenon of land reclamation highlights is no less than a shifting *lebensraum*: a legal expansion of the territorial space through which a sovereign may govern through the slow violence of terraforming – not simply appropriating land but conjuring it from the water.

The viscosity of coastal borders augments a key insight. Far from a finite and unchanging resource, territory in its modern conception is, as Stuart Elden argues, a particular *technology* of sovereignty rather than an objective fact: a "distinctive mode of social/spatial organization" that is "historically and geographically limited and dependent, rather than a biological drive or social need."[8] To think of land reclamation as a distinctly new form of appropriation would miss the fact, therefore, that territory has always been a political mode and logic of spatial organization. Territories have never been the fixed, immobile delineation of the physical extent of a state's bounded jurisdiction.[9] In the rearrangement of borders and states, "new" territory has always come from somewhere else. This is as true of terraformed land as it is of its older precedents: colonialism and military conquest. Often left unexamined in an emerging scholarly interest in large-scale geoengineering projects, then, is the question of what is removed or lost in these acts of sovereign making. It is easy to forget in the spectacular emergence of "new" landmasses that these very acts of movement and creation also require their shadowy, barely traceable other: concomitant acts of extraction, erasure, and dispossession.

In pausing over the term "reclamation" for a while, one might recognize that dubbing an act of terraforming as "reclamation" is a misnomer. In its nominal form, "reclamation" turns a process of territorial acquisition into an abstract act of restoration or return, conveying a process in which the state is not acquiring but rather claiming something *back* through the reassertion of a right, retrieving something that was once one's own. This works as a fiction on two registers. First, it presupposes that the coastal sea itself acts somewhat as *aqua nullius*, "empty" space that has no history or value except to be turned into the property of the state, with the corollary that reclamation is coextensive with an active dispossession from elsewhere. This naturalizes a thoroughly human process of dispossession as a form of natural right. Second, to name the process a form of "*re*claiming" centres the spatial locus of activity on the site in which land is being created, rather than from where it is being taken away. In the logic of reclamation, a state deserves to procure or cultivate a site for habitation or commerce; few questions are asked about the sites from which material has been extracted, and therefore made increasingly uninhabitable.

Some historical context, then. In 1819, when Sir Stamford Raffles, the colonizer of Singapore, wrote excitedly about his "discovery" of Singapore's potential as an entrepôt for the East India Company, he depicted Singapore as a "fulcrum upon which empire shall thrive."[10] This narrative – that an island's primary function is to serve a transitory role as a nodal point in global trade and shipping – has long been the imagined *raison d'être* of Singapore's existence. The logic of the nation as always in transition and grateful for its colonial past runs deep into its identity – from education focuses that shift each time the ruling party determines the next big industry in which it can gain competitive advantage to tourist marketing strategies that proclaim Singapore the "Gateway to Asia." Today, a gleaming white statue of Raffles stands at the landing site where he first set foot on the island. With one leg planted in front of the other, arms folded, Raffles gazes into the distance at the river's mouth. The plaque below his feet reads: "On this historical site, Sir Stamford Raffles first landed in Singapore on 28 January 1819, and with genius and perception, changed the destiny of Singapore from an obscure fishing village to a great seaport and modern metropolis."

In historical accounts of Raffles' founding of Singapore, the geophysical term "backwater" often accompanies descriptions of Singapore's origins as a fishing village. The narrative is always the same: Singapore's place in the world would have been inconsequential had it remained an undeveloped "backwater," an obscure village; Raffles' arrival changed its "destiny" by turning village into metropolis and backwater into concrete jungle. In these nationalist accounts, I do not read the geological metaphor of the backwater as a coincidence. Prior to its colonization, Singapore's shores were primarily marshland and swamp, providing fertile ground for Indigenous groups, the Orang Laut, to fish for mudskippers. The transformation of this so-called backwater fishing village not only removed their livelihoods but also altered the structure of the earth itself: leaching the water from mud, pulverizing Indigenous ways of life, and dredging particulates and wildlife from the marshes to lay the foundations for a port. The theft, command, and control of lebensraum through terraforming is thus at the very centre of Singapore's story of colonial nationhood and the global trade it would facilitate.

But what is perhaps new about the modern terraforming project is that in transporting sand from neighbouring countries, it quite literally steals territory from them. However, rather than doing this through the art of warfare – through territorial occupation or settler colonialism – this theft of land is practically untraceable: islands that one might once have been able to map with coordinates disappear – or, rather, disintegrate – into fragmentary, fungible particles – sand from a disappearing island in Indonesia is practically indistinguishable from sand from a seabed off the coast of the Philippines. In this sense, the national imaginary in which Singapore sustains an articulation of itself as an ever-expanding modern, thriving centre of trade and digital life literally requires a theft of territory, a theft of land – war by other means, war by means of terraforming.

I'm haunted most by the fact that sand is being leached from the very countries from which Singapore extracts most of its foreign labour. It is Cambodians, Burmese, Bangladeshis, and Indonesians upon whose construction work the state relies to build the terraformed habitats in which Singaporeans live. These are the very people whose communities live on or around disappearing islands and depleting marine life. In some Indonesian islands such as Riau, fishing communities have reported that incomes have plummeted as much as 89 percent since the sand trade began.[11] Experts have likewise reported extensive damage to coral reefs, exacerbated coastline erosion, and the destruction of ocean environments that will take decades to be restored.[12]

There is morbid irony in noting these environmental impacts of extraction: the very anthropogenic changes caused by such forms of extraction have become part of Singapore's justification for land reclamation.[13] Officials have cited sea level change as a primary motivation for raising the level of reclaimed seabeds, portraying Singapore as an entropic victim of climate change, even as the sandy bulwarks that ostensibly protect the island from such processes play a key role in exacerbating their effects. In fact, the labourers hired to do the work of such infrastructural development are often precisely those driven from their own communities by such predatory practices of extraction – hired on short-term, contingent, and extremely low-waged contracts to perform highly dangerous work. In this way, the workers charged with increasing the value of Singapore's sovereign and commercial space – by building the highways, condominiums, and business hubs that make Singapore an attractive site for foreign investment – facilitate their own dispensability by constructing the very infrastructure that contributes to the decimation of their lands and the dispossession of their ways of life.

There is a more direct connection between the exploitation of foreign labour and the terraforming of land than I have indexed here. From Hong Kong's port to Dubai's palm-shaped archipelagos, from Macau's casino-jammed Cotai Strip to Singapore's landmark Gardens by the Bay, reclaimed land constitutes a lucrative site of state investment not only because of its ability to expand lebensraum but also because of its commercial value: terraformed land pays hubristic testament to the ability of human hands to remake environments, turning artificially shaped land into spectacles of economic growth. Every towering hotel or palm-shaped bay thus obscures the unevenness by which reclaimed land becomes a fictive commodity: land becomes financialized as real estate at the same time that the labour that builds it is exploited as a cost-saving measure. The contradictions that follow abound. Foreign workers seldom gain access to the glimmering places they help to build, except perhaps to maintain their infrastructure. They are indispensable to Singapore's workforce, but only as surplus populations who can rarely hope to gain citizenship or long-term employment. Meanwhile, the price of terraformed real estate grows as states display it as a marker of sovereign wealth. In this way, the exploitation of labour, the spectacle of sovereignty, and the financialization of real estate go hand in hand.

It is in the disposability of these workers' bodies that I hear echoes of the coolies who once toiled as slaves on the shores of Singapore during the colonial era. There is a national song that begins with the words: "We built this nation, with our hands; the toil of people, from a dozen lands." Those lyrics resurface often when I think about the foreign workers who effectively serve as Singapore's fungible coolie labour today. I wonder: if they had heard that song blasting through speakers during last year's unveiling of the port from their squalid housing quarters in the corners of Singapore, would they have recognized themselves in those words – "we built this nation with our hands" – and the quite literal, material, and violent histories of exploitation, global inequality, and environmental destruction that accompany them.

NOTES

1   Prime Minister Lee Hsien Loong, speech at the opening of Pasir Panjang Terminal Phases 3 and 4, Singapore, June 23, 2015, http://www.pmo.gov.sg/mediacentre/transcript-speech-prime-minister-lee-hsien-loong-opening-pasir-panjang-terminal-phases-3.

2   Kiran Pereira, "Sand Mining: The 'High Volume–Low Value' Paradox," *Aquaknow*, 2012, https://aquaknow.jrc.ec.europa.eu/news/sand-mining-high-volume-low-value-paradox; and UNEP, "Sand, Rarer Than One Thinks," *UNEP Global Environmental Alert Service*, March 2014, http://www.unep.org/pdf/UNEP_GEAS_March_2014.pdf.

3   B. Guerin, "The Shifting Sands of Time – and Singapore," *Asia Times*, July 31, 2003, http://www.atimes.com/atimes/Southeast_Asia/EG31Ae01.html.

4   While the Myanmar Port Authority officially sanctions these dredging and mining operations, numerous reports from Myanmar news outlets state that locals in the Tanintharyi Region now face landslides and erosion due to the digging up of the Dawei River Basin. See, for example, Andrew Loh, "Singapore's Thirst for Sand Again in the News," *Online Citizen*, April 5, 2014, http://www.theonlinecitizen.com/2014/04/spores-thirst-for-sand-again-in-the-news/.

5   Global Witness, *Shifting Sand: How Singapore's Demand for Cambodian Sand Threatens Ecosystems and Undermines Good Governance*, online report, May 2010, https://site-media.globalwitness.org/archive/files/pdfs/shifting_sand_final.pdf.

6   Chris Milton, "The Sand Smugglers," *Foreign Policy*, August 4, 2010, https://foreignpolicy.com/2010/08/04/the-sand-smugglers/.

7   Joshua Comaroff, "Built on Sand: Singapore and the New State of Risk," *Harvard Design Magazine*, 39 (2015), http://www.harvarddesignmagazine.org/issues/39/built-on-sand-singapore-and-the-new-state-of-risk.

8   Stuart Elden, *The Birth of Territory* (Chicago: University of Chicago Press, 2013).

9   For excellent critiques of this logic, still taken as the foundation of sovereignty in international-relations scholarship, see John Agnew, "The Territorial Trap: The Geographical Assumptions of International Relations Theory," *Review of International Political Economy* 1, 1 (1994): 53–80; and Nisha Shah, "The Territorial Trap of the Territorial Trap: Global Transformation and the Problem of the State's Two Territories," *International Political Sociology* 6, 1 (2012): 57–76.

10 Thomas Stamford Raffles, "Report on the Commercial Advantages and Resources Which May Be Expected from the Establishment of a Factory in Singapore," report to Joseph Dart Esquire, Factory Records, Sumatra, 1819, British Library, London, India Office Records, IOR/G/35/50, Board's Collections [F].

11 M. Teguh Surya, "No More Sand Mining!," 2003, http://web.archive.org/ web/20060309183447/http:/www.eng.walhi.or.id/kampanye/tambang/ galianc/no_more/.

12 See, for example, Global Witness, *Shifting Sand*, and UNEP, "Sand, Rarer Than One Thinks."

13 For the negative environmental impacts of sand mining and land reclamation in other contexts, see: Marius Dan Gavriletea, "Environmental Impacts of Sand Exploitation: Analysis of Sand Market," *Sustainability* 9, 1118 (2017): 2–26; M. Naveen Saviour, "Environmental Impact of Soil and Sand Mining: A Review," *International Journal of Science, Environment and Technology,* 1, 3 (2012): 125–34; and Vince Beiser, "Sand Mining: The Global Environmental Crisis You've Probably Never Heard Of," *The Guardian,* 2017, https://www.theguardian.com/cities/2017/feb/27/sand-mining-global -environmental-crisis-never-heard.

Singapore as "Best Home"

Natalie Oswin

IN THE LATE 1990S, following more than three decades of incredible manufacturing-led economic growth, Singapore's competitiveness waned as the geographies of the international division of labour shifted. So its People's Action Party (PAP) – which has governed without interruption throughout the postcolonial period – set about nurturing knowledge-based sectors. By the early 2000s, the transition to a "globalized, entrepreneurial, diversified" economy was well under way, and a spate of articles detailing policy adviser Richard Florida's ideas about creative cities, including his ideas about how tolerance for LGBT persons signals the presence of a strong creative class, appeared in the main local daily newspaper, *The Straits Times* (a media outlet that, like all media outlets in the city-state, is run with strong government ties and oversight).[1]

One representative piece, titled "Making Room for the Three T's," states that "the creative class wants to be where there is a happening scene, a pulsating music and arts environment, and a tolerant and diverse population ... A city needs to focus on getting the right 'people climate.'"[2] Singapore has taken this advice. In 2005, Prime Minister Lee Hsien Loong stated that his government was embracing the "need to remake our city, so that it is vibrant, cosmopolitan and throbbing with energy, with our own distinctive X-factor that makes us stand out from other cities."[3] It has also taken the advice that tolerance of LGBT persons is important for the creative city, with the significant caveat that mere tolerance is enough. That is, efforts to stamp out same-sex sexual activities and discourage LGBT community building, which were common through the late 1990s (efforts such as well-publicized police raids of gay bars and cruising grounds that led to multiple arrests, caning sentences, and the publication of entrapped men's pictures in local newspapers),

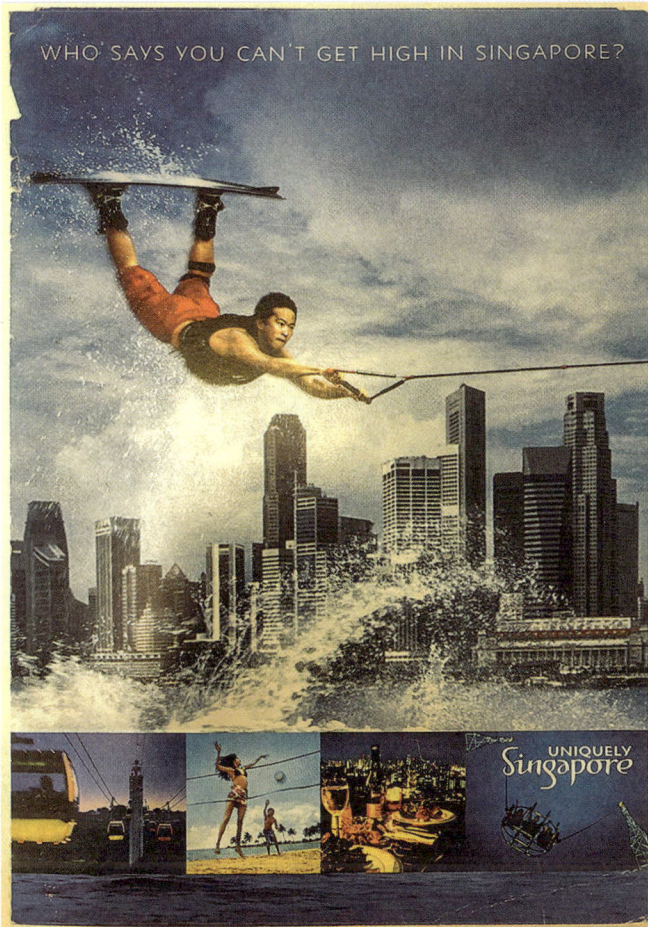

WHO SAYS YOU CAN'T GET HIGH IN SINGAPORE?

UNIQUELY
Singapore

Postcard from Singapore Tourism Board's "Uniquely Singapore" campaign, early 2000s. While highlighting "urban buzz," the tagline alludes to Singapore's hard-line stance on recreational drug sales and consumption. Photograph courtesy of author.

have largely ceased. Yet legislative change has not followed, and LGBT Singaporeans remain second-class citizens in all kinds of ways – not least, a colonial-era penal code statute criminalizing "sodomy" remains on the books.

The fight for LGBT rights and dignity thus continues in the creative city of Singapore and, thankfully, a strong LGBT movement has emerged to take up this cause. In this essay, I acknowledge the dubious ways that LGBT issues have been enfolded into Singapore's urban and national development plans. Moving deliberately away from the literal referent of sexual-identity politics, I situate LGBT issues within a broad field of power relations to analyze the city-state as not a hetero*sexual* space but a hetero*normative* space.[4] I advance, in other words, a queer reading of

Singapore that sheds light not only on LGBT lives but also on other "others" who do not fit into the government's notions of respectable domesticity and proper family. I contextualize digital Singapore by focusing on its analog roots, its embodied politics of population.

To create the right "people climate" for the creative city, the PAP made dramatic changes to the city-state's urban landscape, including infrastructural investments in universities, arts centres, and biotechnology hubs; the development of "lifestyle quarters" such as One North and South Beach; the revitalization of leisure areas such as the Singapore River, Dempsey Road, and Sentosa; the creation of green spaces and parks throughout the island and the building of a new downtown on reclaimed land; the upgrading of public-housing complexes to offer housing options for the upwardly mobile; and the expansion of enclave urbanism in the form of private-housing developments for the local and global elite. Profound changes to the composition of Singapore's population have also ensued, changes that were first signalled in Prime Minister Lee's 2006 National Day Rally speech: "If we want our economy to grow, if we want to be strong internationally, then we need a growing population and not just numbers but also talents in every field in Singapore ... There are things which we can do as a government in order to open our doors and bring immigrants in. But more importantly as a society, we as Singaporeans, each one of us, we have to welcome immigrants, welcome new immigrants."[5]

Singapore, as a small island nation without a natural resource base, depends heavily on human resources for economic growth and has supplemented its local population with foreign migrants for most of its postcolonial history. But before 2006, these migration flows had generally been regulated as temporary. The turn to welcoming new immigrants by encouraging some of them to become naturalized Singapore citizens was therefore a significant departure.

This was not the government's first choice. Until 2006, the PAP had relentlessly attempted to reproduce the population through natural increase, and it still constantly exhorts Singaporeans to marry and procreate. This fact is often joked about – for example, in a 2012 Mentos commercial, Singaporeans are implored to have sex on National Day with the slogan "Let's Not Watch Fireworks, Let's Make 'Em Instead." But the government is by no means kidding around. As is well known, the city-state's extraordinary postcolonial socioeconomic development was achieved through colossal feats of social engineering. The "up-grading" of the family was at the core of these efforts. For much of its colonial period, Singapore was an entrepôt with a majority population of single male migrant workers. Family formation and unification were encouraged only in the last decades of colonial rule, when, in line with shifting ideas about the relationship between population and labour in Britain and other parts of its empire, "it seemed cheaper to nurture children to adulthood than to import adult workers."[6]

But it was the postcolonial government that would balance the sex ratio and get serious about population policy. In local social-reform literature written at the time of independence, the existence of a large percentage of atypical households (that is, households comprising singles, large families, or multiple families) is cast as a key impediment to development. Thus, the PAP began to put "proper" Singaporean families in place by making family planning a high priority. As the first post-independence minister for health stated: "Family Planning is ... a matter of national importance and indeed, one of urgency for us. Our best chances for survival in an independent Singapore is stress on quality

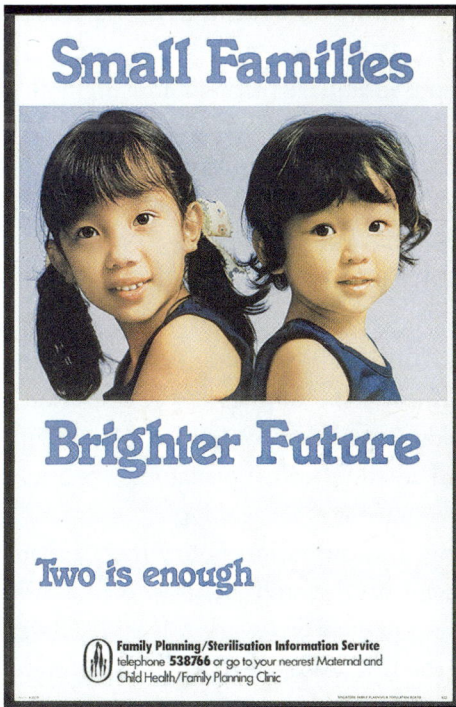

Family Planning and Population Board campaign poster, 1978. Courtesy of the National Archives of Singapore.

and not on quantity."[7] Voluntary sterilization was legalized in 1969, and 112,568 sterilizations were performed between 1969 and 1988. Induced abortion was also legalized in 1969, and no fewer than 288,666 abortions were performed in the same period. Further, a range of tax policies that disincentivized "higher order births" (that is, having more than two or three children) were rolled out in 1973.[8]

Alongside efforts to limit family size, migration policy was used to create the PAP's desired Singaporean population. Since 1933, various Aliens Ordinances had effectively limited immigration to reuniting families in the late colonial period. In the initial independence period, strict controls on the import of foreign workers were put in place as the new city-state "buckled down to the task of nation building" and worked at creating a population of families in part by defining "fixed categories to incorporate citizens of the nation on the one hand and exclude others as noncitizens, or 'aliens' on the other."[9] As industrialization gathered steam in the 1970s, immigration controls were relaxed and inflows of unskilled labourers – or what the Singapore government refers to as foreign workers – from countries such as the Philippines, Indonesia, Thailand, India, Bangladesh, and Sri Lanka were permitted. But the number of immigrants from these countries was small through this decade, their work permits temporary.

The Singapore family was also put into place through housing policy. The Housing Development Board (HDB), which was formed by the PAP in 1960 to provide "homes for the people," has progressively covered the island with modernist high-rise apartment blocks in which more than 85 percent of the current population resides. Its tenancy requirements have been consistent since the HDB's founding. To purchase a flat, the applicant must be at least twenty-one years of age and "form a proper family nucleus," which is defined as follows: the applicant and fiancé; the applicant, spouse, and children (if any); the applicant, the applicant's parents, and siblings (if any); if widowed or divorced, the applicant and children under the applicant's legal custody; and, if orphaned, the applicant and unmarried siblings.[10]

Thus, using the family, housing, and migration policy mechanisms at its disposal, and getting its message out through extensive media campaigns, the PAP "created an atmosphere of crisis and identified large families as an imminent threat to the limited resources of the city-state" in the early years of its rule.[11] Then, because the antinatalist policies of the 1960s and 1970s had been so effective that the fertility rate fell below the replacement level by the early 1980s, the city-state became the first newly industrialized country to adopt pronatalist policies. Immediately upon making the shift to pronatalism, the eugenicist logic that animated the antinatalist policies – its emphasis on creating a "quality" population by limiting childbirths among low-income women – was brought explicitly to the fore in the "Great Marriage Debate." In 1983, Prime Minister Lee Kuan Yew reprimanded the nation's mothers for failing to reproduce themselves along race and class lines in appropriate numbers. He encouraged those he referred to as graduate mothers – that is, women with university educations and belonging disproportionately to the Chinese community – to replace themselves, and the government discouraged poorly educated women of generally Malay or Indian backgrounds from reproducing too freely.

Incentives for the former group included "generous tax breaks, medical insurance privileges, and admission for their children to the best schools in the country" while disincentives for the latter group came in the form of cash awards "to restrict their childbearing to two children, after which they would 'volunteer' themselves for tubal ligation."[12] There was much backlash, particularly from the women's movement.[13] Some of Lee Kuan Yew's more contentious claims – such as explicit linkages between race and class and fitness for parenthood, and a call to bring back polygamy for those members of the population who were most "vital" – were tempered. But the general rationale of encouraging the "right" people to populate the city-state remained as tertiary-educated women were singled out as particular problems and labelled as overly career-minded and too choosy in their choice of partners. From the state's eugenicist perspective, "the loss of their progeny was a loss to the nation's talent pool."[14]

In 1986, the Population Policy Unit was founded to enact policies that would encourage the educated and economic elite to have many children; it used a combination of incentives and disincentives to encourage everyone else to have fewer. Over the last thirty years, those encouraged to have more children have been able to access tax breaks, housing subsidies, baby bonuses, paid maternity leaves, and free medical services such as discounted delivery room fees and publicly provided assisted reproductive treatments – an array of incentives deliberately held out of reach of lower-income families and divorced or single parents.

Disincentives for the latter group include the perpetuation of sterilization schemes, subsidized abortion access, legal prohibition from using assisted reproductive technologies, less government paid maternity leave time, and so on.[15] In addition, in 1984, the PAP set up the Social Development Unit, a national matchmaking agency aimed at assisting university graduates to find partners. The Social Development Service was subsequently formed to assist those with lower educational qualifications and, in 2009, as anxieties over the persistent below-replacement fertility level deepened, the Social Development Unit and the Social Development Service were merged to form the Social Development Network in a bid to give members a wider pool from which to choose. The state also promotes proper family formation through frequent public education efforts launched by the Public Education Committee on Family. Its initiatives include marriage preparation workshops and guidebooks, radio and television programming on the value of family, and campaigns such as 2003's "Romancing Singapore." As Kenneth Paul Tan notes, the campaign "aimed to celebrate 'life, love and relationships' by encouraging heterosexual couples, caught up in fast-paced lifestyles, to 'be more expressive with their partners at all times, not just on special occasions.'"[16]

On top of dealing with the demands of "fast-paced" lifestyles in a global city, the Singapore government faced another major challenge in maintaining its optimum population – outward migration flows. "Elites" at the top of its "gene pool" are cast as a big part of the "talent" pool that the city-state needs to keep its creative economy afloat. Thus, the PAP puts tremendous resources into casting Singapore as the "best home" for talent. In the 2000s, it set up the Overseas Singaporean Unit and the National Population and Talent Division of the Prime Minister's Office with mandates to find ways to maximize Singaporean talent, either overseas or in the city-state, for economic growth.[17] It depends on its media and communications arms to get the message out, and National Day speeches and rallies stressed the theme of Singapore as "best home" without fail for well over a decade.

So this is the context in which Singapore has relaxed its immigration policies. While the dominant discourse of the government-run mainstream press touts the necessity of this policy shift, worries over employment prospects for nonimmigrant Singaporeans and the nature of Singapore identity in this new cosmopolis are often articulated in the

An ad for one of Singapore's many new condo developments marketed to local and foreign "talent." Photograph by author.

independent online press. What is far too rarely remarked upon, however, is the fact that the state does not welcome just any immigrants. Rather, it maintains its emphasis on creating a quality population by concertedly encouraging the naturalization of *suitable* immigrants. The potential new immigrants are "foreign talent" or "quality people," in the terms used in the 2010 report of the "ESC subcommittee on Making Singapore a Leading Global City."[18] An information pack produced by the Immigration and Checkpoints Authority titled *Embrace Singapore, Where You Belong!* lays out this invitation clearly. It states: "In the embracing spirit of our forebears, we welcome you to our shores, to be part of our country's vitality ... Together let's work and contribute to our continued prosperity and shape our future here – our Home." It implores: "Share your talents. Build a nation."[19]

This invitation to naturalization is an invitation to join the national family, literally as well as figuratively. The fact that Singapore's government now considers these migrants an essential factor of the city-state's social reproduction is expressed as follows in the same ICA information

Serangoon Gardens' foreign-worker dormitory. Photograph by author.

pack: "Many of you have even brought family members to Singapore to help you better adjust to living here ... You and your family members have benefited from what Singapore has to offer, just as Singapore has progressed and prospered with your labour and contributions. It is time to take a step further and become a part of the Singapore family."[20]

A dedicated section titled "Dear Family Members" further details "what Singapore has to offer to foreigners and their families." The listed "advantages of living, working and learning in Singapore" include the following: "a standard of living rated amongst the highest in Asia Pacific and comparable to many first world countries"; a "comfortable pace of life, quality education and health care"; "good transport and utilities infrastructure"; "and the feeling of safety and security in a country where crime rates are one of the lowest in the world." All of these fruits of Singapore's socioeconomic development to date are offered to "foreign talent" so that they will "help her [Singapore] remain continually relevant to the dynamic and changing environments on the world stage."[21]

"Foreign talent" – particularly people who will form "proper" families and help reproduce the nation – are cast as harbingers of a bright future and offered a secure place within the city-state. This is significant not least because inviting foreign talent into Singapore's developmental future opens the carefully crafted Singaporean family up to new members. At the same time, however, much remains the same, for this family's borders are being reinforced to keep another migrant population – and a much more numerically significant one – out of public life and beyond the pale of public participation. While government discourse emphasizes bringing "talent" to Singapore, its efforts to become a leading creative global city have been accompanied by a much larger surge of "foreign workers" to the city-state. In 2009, Singapore's foreign population numbered 1.2 million, approximately a quarter of its total population. Within this group, only around 170,000 were expatriate professionals (that is, "foreign talent") while the remainder were low-wage foreign workers, largely employed in the domestic service and construction sectors.[22]

In contrast to foreign talent, there are no glossy brochures inviting foreign workers into the national family. Rather, they are rendered permanently transient and alien through a range of policies that regulate their time in Singapore. Crucially, this exclusion is enacted fundamentally through exclusion from the institution of the family and the sphere of social reproduction. As the Employment of Foreign Manpower Act states: "The foreign employee shall not go through any form of marriage or apply to marry under any law, religion, custom or usage with a Singapore Citizen or Permanent Resident in or outside Singapore ... If the foreign employee is a female foreign employee, the foreign employee shall not become pregnant or deliver any child in Singapore during and after the validity of her work permit ... The foreign employee shall not be involved in any illegal, immoral or undesirable activities, including breaking up families in Singapore."[23] In other words, in a city-state that has long been obsessed with reproducing its position at the front of the global city pack, foreign workers have no future.

While Singapore's progress narrative rolls on, and the state works at making it the best home for some, foreign workers are rendered abject and alien as they are deliberately made to exist outside the sphere of intimacy, love, and familial connection – and thus outside the nation. The eugenicist logic that once overtly guided Singapore's population policies is commonly narrated as part of its past. But in this "best home," it is clearly alive and well, and still keeping vast numbers of people on the margins.

NOTES

1 Lee Hsien Loong, *New Challenges, Fresh Goals: Towards a Dynamic Global City* (Singapore: Ministry of Trade and Industry, 2003). For an overview of state-based media control in Singapore, see Cherian George, *Freedom from the Press: Journalism and State Power in Singapore* (Singapore: National University of Singapore Press, 2012).

2 "Making Room for the Three T's," *Straits Times,* July 14, 2002.

3 Lee Hsien Loong, National Day message, August 20, 2006, University Cultural Centre, Singapore, http://www.nas.gov.sg/archivesonline/speeches/record-details/7e69abe9-115d-11e3-83d5-0050568939ad.

4 Heteronormativity is the geographically and historically specific coincidence of race, class, gender, and sexual norms. Thus, it reaches far beyond sexual- and gender-identity struggles to shape familial and intimate relationships, domestic norms, migration flows, national identities, and more.

5 Ibid.

6 Lenore Manderson, *Sickness and the State: Health and Illness in Colonial Malaya, 1870–1940* (New York: Cambridge University Press, 1996), 203.

7 Singapore Family Planning and Population Board, "Family Planning: A Series of 12 Papers on Family Planning from Dec. 1965 to Dec. 1967," Singapore, Ministry of Health, Public Health Division, 1968.

8 David Drakakis-Smith, Elspeth Graham, Peggy Teo, and Giok Ling Ooi, "Singapore: Reversing the Demographic Transition to Meet Labour Needs," *Scottish Geographical Magazine* 109, 3 (1993): 152–63.

9 Brenda S.A. Yeoh, "Bifurcated Labour: The Unequal Incorporation of Transmigrants in Singapore," *Tijdschrift voor Economische en Sociale Geografie* 97, 1 (2006): 26.

10 Unmarried, divorced, or widowed individuals and single parents become eligible to buy HDB flats after the age of thirty-five. But even then, they are only eligible to purchase resale, rather than new, flats, and they receive a much smaller subsidy than family applicants. See the HDB's website for full statement of the eligibility criteria for various flat-ownership and -sublet schemes.

11 Janet W. Salaff, *State and Family in Singapore: Restructuring a Developing Society* (Ithaca, NY: Cornell University Press, 1998).

12 Geraldine Heng and Janadas Devan, "State Fatherhood: The Politics of Nationalism, Sexuality, and Race in Singapore," in *Bewitching Women, Pious Men: Gender and Body Politics in Southeast Asia,* ed. Aihwa Ong and Michael Peletz (Berkeley: University of California Press, 1995), 195–215.

13 For an account of the clash between activists and the PAP on this issue and its importance in the history of the Singapore feminist movement, see Lenore Lyons, *A State of Ambivalence: The Feminist Movement in Singapore* (Boston: Brill, 2004).

14 You Yenn Teo, "Shaping the Singapore Family, Producing the State and Society," *Economy and Society* 39, 3 (2010): 339.

15 Lim Keak Cheng, "Post-Independence Population Planning and Social Development in Singapore," *GeoJournal* 18, 2 (1989): 163–74; Drakakis-Smith et al., "Singapore"; Shirley Hsiao-Li Sun, *Population Policy and Reproduction in Singapore* (London: Routledge, 2012); and Teresa Wong, Brenda S.A. Yeoh, Elspeth Graham, and Peggy Teo, "Spaces of Silence: Single Parenthood and the 'Normal Family' in Singapore," *Population, Space and Place* 10, 1 (2004): 43–58.

16 Kenneth Paul Tan, "Sexing up Singapore," *International Journal of Cultural Studies* 6, 4 (2003): 403–23.

17 Elaine L.E. Ho, "Constituting Citizenship through the Emotions: Singaporean Transmigrants in London," *Annals of the Association of American Geographers* 99, 4 (2009): 788–804.

18 "ESC Subcommittee on Making Singapore a Leading Global City," report, http://www.mof.gov.sg/Portals/0/MOF%20For/Businesses/ESC%20Recommendations/Subcommittee%20on%20Making%20Singapore%20a%20Leading%20Global%20City.pdf.

19 *Embrace Singapore, Where You Belong!*, accessed online July 7, 2016, at https://www.ica.gov.sg/news_details.aspx?nid=2913.

20 Ibid. For additional explicit statements of the government's "talent-centric" approach, see the website of the National Population and Talent Division, and specifically its Economic Strategies Committee.

21 Ibid.

22 Grace Baey, "Borders and the Exclusion of Migrant Bodies in Singapore's Global City-State" (master's thesis, Queens University, 2010).

23 *Employment of Foreign Manpower Act*, c 91A.

Not another Cinderella Story

Symon James-Wilson

> Cinderella never asked for a prince.
> She asked for a night off and a dress.
>
> – Kiera Cass

BEFORE ANIMATED FILMS and computer-generated imagery, there was a Cinderella story unfettered by wizarding fairy godmothers and magic pumpkins. Unlike Disney's 1950 and 2015 adaptations, the Brothers Grimm version of this classic tale challenges the predictable victimization of Cinderella as a persecuted heroine awaiting rescue. Describing a young woman's artful evasions of an entitled admirer (she jumps into chicken coops, climbs pear trees, and sprints down long flights of stairs to escape the prince's advances), Jacob and Wilhelm Grimm portray Cinderella as clever, resourceful, and unyielding. Hardly a damsel in distress, she actively refuses to be "saved" from the commonness of her daily life by external forces. As Kiera Cass's quote so poignantly suggests, neither the wave of a wand nor Prince Charming's kiss can take credit for her empowerment. Cinderella's independent pursuit of her basic human rights (a night off from work and proper clothing) is the central catalyst for her metamorphosis from household fixture to author of her own story.

Far from the dazzling ballroom of the king's castle, a more careful reading of this fiction unveils its deeper truth. Cinderella's brightest moment is not embellished with silver and gold. Instead, it rises from the smoke and ashes of a kitchen stove. Why are many of our most transformative instances of self-advocacy and resistance overlooked? What would happen if we examined these moments with intention?

Creative reimaginations of this timeless tale continue to unfold in contemporary global cities. Provocatively called "global Cinderellas" by Taiwanese sociologist Pei-Chia Lan, the 8.5 million female migrant domestic workers around the world today have appeared in many popular and academic narratives as victims of globalization.[1] Focusing on the precarity of their temporary legal and employment statuses in destination countries and the "brokenness" of the families they leave behind, the dominant discourse on gendered labour migration historically trivialized domestic workers as submissive pawns in the concealed manoeuvrings of late capitalism's uneven development.[2]

Rather than acknowledge a long history of racialized women disrupting patriarchal economics shackled to the idea of a sole male breadwinner, popular discourses continue to present female migrant domestic workers of the new millennium through the conventional archetypes of the docile maid, sexualized object, and vulnerable "Third World" woman.[3] The rhetorical emphasis on exploitability lends itself to the commodification of female migrant domestic workers as (im)mobile objects of multinational social policy and modes of governance. The dichotomized language that represents countries as exclusively migrant-sending or migrant-receiving reinforces this invention that gendered labour migration is a linear and uncomplicated economic transaction.[4] The reduction of migrant domestic workers to orderly care packages, assembled and delivered to cure the symptoms of dismantled welfare states in postindustrial countries, is a strategic and geographic project. It is fundamentally aimed at legitimizing the dehumanization of migrant domestic workers through the supply-and-demand logics of "necessity," powerfully suggesting that the health and well-being of wealthy urban societies in the global north and south are the ends that justify the means.[5] This traditional mythscape of gendered labour migration ultimately distracts from these workers' immensely pioneering efforts to not only provide economically for their immediate and extended relatives amid persistent global inequality but also to simultaneously reproduce (through cooking, cleaning, and caretaking) dynamic societies around the world.

The overgeneralized characterization of these workers as the damsels of global distress is, fortunately, not the only interpretation of gendered labour migration. In global cities such as Hong Kong, Taiwan, and Singapore, there have been considerable efforts to recentre human agency in depictions of not-so-fairy-tale-like urban settings.[6] Suggesting that idyllic yet dystopian cityscapes generate more complex power geometries than clear winners and losers, much of this discussion has focused on how the technological advancements of the digital age have transformed female domestic workers' personal and work experiences of transnational migration. In Singapore, in particular, advancements in the availability and affordability of information and communication technologies (ICTs) since the early 1990s have had a remarkable impact on the small island city-state's global status as a leading "wired" metropolis.[7] In this world-renowned hub of finance, transit, and information, more than 231,000 migrant domestic workers are said to have gained considerable benefit from working in a country where technology is so ubiquitous.[8] Numerous studies on these workers' ICT usage have linked access to mobile phones, video calling (e.g., Skype and Facetime), and social-media platforms (e.g., Facebook and Twitter) with greater social capital, stronger personal networks, reduced stress, and enhanced emotional well-being.[9] Delivering passionate accounts of letter writing and public phone booths being rendered almost completely obsolete within less than two decades, researchers and journalists alike have celebrated the expansion of global telecommunications as a transcendent revolution. In this context, the "Cinderella with a mobile phone" is confidently portrayed as an enterprising and tech-savvy global citizen. Digital technologies allow female migrant domestic workers new opportunities for transnational mothering (by allowing them to keep in contact with their children and spouses overseas) and to make new claims on public space (e.g., by frequenting shopping malls to purchase 3G data packages).[10]

However, even this arguably more progressive narrative is not without its own cautionary tales. While it is undeniable that modern communication technologies have enhanced female migrants' capacity to compress the distance between their working lives "here" and family lives "over there," the proliferation of ICTs has nevertheless failed to drastically overhaul prevailing paradigms. As is suggested by American sociologist Rhacel Parreñas in her seminal ethnographic study of Filipina migrant domestic workers, *Servants of Globalization*, applauding the "technological management of distance" between migrant mothers and families left behind as a small-scale form of worker agency ultimately diverts attention from the unchanging realities of world financial crises, compounded indebtedness, and official policies of forced migration that continue to have profound daily effects in postcolonial countries such as Indonesia and the Philippines.[11] Likewise, this discussion has done relatively little to challenge neoliberal globalization's ongoing deregulation of mobile goods (e.g., cell phones) and more stringent restriction of mobile people (e.g., migrant domestic workers) in global cities such as Singapore.[12] Although redirecting focus away from accounts of modern-day slavery towards a more vibrant discussion of technological emancipation has made important contributions, "techno-euphoria" often obscures the everyday consequences of ICTs (mis)uses.[13] From employment agencies confiscating workers' communication devices upon their arrival in Singapore to employers using GPS tracking embedded in smartphones to monitor their live-in employees' every movement, advancements in communication technologies have reinforced patriarchal power relations as much as they have troubled them.[14] By framing migrant workers' use of modern communication technologies through a conceptual framework that emphasizes displacement and a nostalgic yearning for a connection to "home," this conversation has often reinforced the very character tropes of female migrant domestic workers that it initially set out to counter.

While descriptions of women spending their Sundays tracking the best sales on prepaid SIM (subscriber identity module) cards in Singaporean shopping centres present female migrant domestic workers as economical and astute, they seldom bring into question the underlying unfairness of requiring sponsorship for access to the more favourable contract phone plans, which most women on two-year work permits cannot obtain from their employers.[15] Reducing multidimensional urban protagonists to static entities that can be technologically managed and manipulated, this framework presents female migrant domestic workers as only partially included within Singaporean society. While these women are strategically integrated into the country's material consumption, their creative social productions are regularly left unacknowledged.

In the winter and summer of 2015, I was given the opportunity to work as a research assistant on a pilot exploration of educational programs for female migrant domestic workers in the Philippines, Indonesia, and Singapore. Adopting both a pedagogical and geographic lens, my project focused on identifying how migrant women's lived experiences are sociospatially constructed and contested through education systems. Engaging with a variety of government, NGO, labour union, private sector, and grassroots programs in each country, I found that many educational initiatives were in fact not narrowly socializing migrants into a culture of subservience, as the prevailing discourse suggested, but were instead creating transformative and empowering experiences that encouraged worker agency.[16]

One Sunday morning in January 2015, I witnessed the transformation of math and science classrooms into beauty salons, dressmaking workshops, and aromatherapy parlors at Singapore's ISS International School. Instantly, I was captivated by the creativity of the Singaporean charity Humanitarian Organization for Migration Economics' (HOME) vocational-training program for female migrant domestic workers. Aptly named "HOME Academy," the program sees approximately five hundred women graduate from skills-based vocational courses held every other Sunday over the course of six months at three campus locations.[17] With classes ranging from baking, guitar, and advanced cosmetology to English-language instruction, financial education, and Javanese bridal styling, the academy's biweekly programing appeals to a wide range of student interests.[18] Additionally, HOME organizes a variety of supplementary workshops on Singaporean employment law, human rights, sexual health, and international conventions pertaining to migrant workers.[19] As one participant explained to me, for many women, HOME Academy represents a "once in a lifetime opportunity." Since 2009, more than five thousand female domestic workers have benefitted from its engaging initiatives.[20]

Shadowing facilitators and interviewing students on three separate visits, I was greatly intrigued by HOME Academy's affinity for traditional communication technologies in the context of the omnipresence of Singapore's expansive technosphere. I conceptualize traditional communication technologies as older information-sharing methods such as flyers, photography, radio, word of mouth, and face-to-face interaction. By constructing practical information pamphlets, distributing HOME Academy biannual yearbooks (filled with full-colour pictures of each course and its participants) at graduation ceremonies, and coming together for special celebrations, female migrant domestic workers created original forms of belonging and knowledge production without digital intervention. At all three campuses, using relatively simple tools and techniques, women had invaluable opportunities to contribute personal stories, share news, offer assistance, and build comradery with others who shared the common bond of a particular lived experience. In these educational spaces, the meaning of transnational community was no longer exclusive to ICT-supported contact with family overseas. Instead, it expanded to include a multilingual, multicultural, and multinational sisterhood of Filipina, Indonesian, Indian, and Burmese women who were able to communicate their mutual understanding of transnational labour migration's intimate daily realities through media such as drama, song, dance, and the spoken and written word.

Supporting domestic workers to develop confidence in their talents and to value their experiential knowledge, HOME Academy recognizes the power of traditional communication methods for fostering self-advocacy and creative problem solving in the face of work- and nonwork-related challenges. The vast majority of course instructors and the principals of all three campuses were domestic workers themselves. By adopting an organizational model that places migrant women at the heart of its organizational infrastructure, HOME Academy is a powerful site for the types of emancipatory pedagogy that can create spaces for women to be viewed as – and, in fact, to become – self-empowered protagonists instead of victims who are simply acted upon. Notably, HOME Academy's deliberate curricular and pedagogical choices convey a carefully nuanced understanding of both education and technology as infrastructures that have the potential to dislocate or re-entrench hegemonic systems of power.

While HOME utilizes digital ICTs such as Facebook and Twitter to advertise upcoming events, repost relevant news articles, and provide virtual outreach to migrant workers in need of assistance (often via direct message or chat windows), its most significant and long-standing advocacy work relies on traditional communication technologies such as face-to-face consultations with government stakeholders and formal letters of appeal to major Singaporean policy makers.[21] These "old" techniques proved to be indispensable during a challenging multiyear campaign to have the Ministry of Manpower mandate a weekly rest day for migrant domestic workers (which was finally passed into legislation on January 1, 2013).[22] Additionally, the continued success of its walk-in Help Desk, paralegal counselling and conflict-mediation services, and shelter for women workers rely on traditional methods.[23]

To gain a deeper understanding of why digital communication mechanisms are not always the preferred vehicle for creating change, I returned to HOME Academy to observe how participants and facilitators in educational settings made decisions about new and old technologies on a local scale. Every March, HOME hosts a talent pageant in celebration of International Women's Day. The competition has prizes in dancing, drama, and singing and an array of coveted titles (including "Miss Catwalk," "Miss Global Wear," and "Miss Congeniality"), and nearly one hundred domestic workers take part in the qualifying, semi-final, and final rounds each year.[24] After speaking to two of the 2015 talent pageant organizers, both domestic workers themselves, the transformative impact of HOME's "bridging" approach to the use of traditional and digital communication technologies in their programs was clear.

At the time of our interview, the first pageant organizer I spoke with had worked as a domestic worker in Singapore for fifteen years. Having been involved with HOME as a volunteer at the Help Desk, she said that taking 130 pageant candidates under her wing for that year's show was unquestionably her biggest assignment yet. She described herself as being aware of her talents but too shy to show them off before becoming involved with the pageant. Given the opportunity to draw from her experiences as part of a dance group at her high school in the Philippines and as a choreographer for Japanese entertainers, the pageant helped her cultivate effective team leadership skills and gain the confidence to self-advocate in her workplace. She described the relationship between pageant contestants and their coaches to be "like a family": they encouraged each other to share their opinions, contribute to group decision making, and join in regular potluck lunches using traditional forms of communication technology.

The second pageant organizer I interviewed corroborated many of the first interviewee's main points, but offered her own distinct wisdom. Although she had only worked as a domestic worker for eight years, she already had a year of HOME pageant coordinating experience in her toolkit. When it came to her favourite things about coaching, encouraging the ladies to try on different outfits and wear makeup, and building friendships through singing and dancing, were clearly at the top of her list. She spoke of bridging the gap between traditional and digital communication technologies in her teaching by encouraging contestants to take and upload photos to Facebook, and she said she even participated in the occasional group selfie herself. Recalling how lonely she had been during her first two-year employment contract in Singapore, she expressed how important it was to her and the other participants to foster social networks. Central to this goal, in her view, was the development of strong communication skills, whether that took the form of writing and delivering public speeches at a pageant or articulating creativity on social media platforms.

Communication technology has developed and will continue to develop in tandem with globally shifting political, social, and economic paradigms. Too often overlooked in the digital discourse on migrant agency, initiatives such as HOME Academy and the HOME talent pageant remind us that traditional communication technologies remain critical in the creation of self-empowering experiences for female domestic workers in Singapore. Further research on the methods and rationale by which social actors mix old and new technological media and strategies to the greatest effectiveness could stimulate considerable productive tension. In particular, deeper consideration of how blended communication strategies operate within educational spaces could inspire new approaches that interrupt and contest prevailing victimization narratives.

NOTES

1 Pei-Chia Lan, *Global Cinderellas: Migrant Domestics and Newly Rich Employers in Taiwan* (Durham, NC: Duke University Press, 2006); and International Labour Organization, *ILO Global Estimates on Migrant Workers: Results and Methodology, Special Focus on Migrant Domestic Workers* (Geneva: International Labour Organization, 2015), xiii.
2 Rhacel Salazar Parreñas, *Servants of Globalization: Migration and Domestic Work,* 2nd ed. (Stanford, CA: Stanford University Press, 2015).
3 Barbara Ehrenreich and Arlie Russell Hochschild, eds., *Global Woman: Nannies, Maids, and Sex Workers in the New Economy* (New York: Metropolitan Books/Henry Holt, 2004).
4 Louka T. Katseli, Robert E.B. Lucas, and Theodora Xenogiani, *Effects of Migration on Sending Countries: What Do We Know?*, OECD Development Centre No. 250 (Paris: Organization for Economic Development and Cooperation, 2006).
5 Franca van Hooren, "When Families Need Migrants: The Exceptional Position of Migrant Domestic Workers and Care Assistants in Italian Immigration Policy," *Bulletin of Italian Politics* 2, 2 (2010): 21–38.

6 Nicole Constable, "Migrant Workers and Many States of Protest in Hong Kong," *Critical Asian Studies* 41, 1 (2009): 143–64; Lan, *Global Cinderellas*; Brenda S.A. Yeoh and Shirlena Huang, "'Home' and 'Away': Foreign Domestic Workers and Negotiations for Diasporic Identity in Singapore," *Women's Studies International Forum* 23, 4 (2000): 413–29.

7 Eric C. Thompson, "Mobile Phones, Communities, and Social Networks among Foreign Workers in Singapore," *Global Networks* 9, 3 (2009): 367.

8 Ministry of Manpower, "Foreign Workforce Numbers," data file, 2015, http://www.mom.gov.sg/documents-and-publications/foreign-workforce -numbers.

9 Arul Chib, Holley A. Wilkin, and Sri Ranjini Mei Hua, "International Migrant Workers' Use of Mobile Phones to Seek Social Support in Singapore," *Information Technologies and International Development* 9, 4 (2013): 19–34; and Humanitarian Organization for Migration Economics (HOME), *Home Sweet Home? Work, Life and Well-Being of Foreign Domestic Workers in Singapore*, research report, 2015, https://idwfed.org/en/resources/home -sweet-home-work-life-and-well-being-of-foreign-domestic-workers-in -singapore/@@display-file/attachment_1.

10 Arul Chib, Shelly Malik, Rajiv George Aricat, and Siti Zubeidah Kadir, "Migrant Mothering and Mobile Phones: Negotiations of Transnational Identity," *Mobile Media and Communication* 2, 1 (2014): 73–93; Lan, *Global Cinderellas*; and Geraldine Pratt, *Families Apart: Migrant Mother and the Conflicts of Labor and Love* (Minneapolis: University of Minnesota Press, 2012).

11 Parreñas, *Servants of Globalization*, 75, 99; and Robyn Magalit Rodriguez, *Migrants for Export: How the Philippine State Brokers Labour to the World* (Minneapolis: University of Minnesota Press, 2010).

12 Juanita Elias, "Gendered Political Economy and the Politics of Migrant Worker Rights: The View from Southeast Asia," *Australian Journal of International Affairs* 64, 1 (2010): 79.

13 Cara Wallis, "Mobile Phones without Guarantee: The Promises of Technology and the Contingencies of Culture," *New Media and Society* 13, 3 (2011): 472.

14 Ibid.; and Thompson, "Mobile Phones."

15 Thompson, "Mobile Phones"; and Brenda S.A. Yeoh and Shirlena Huang, "Negotiating Public Space: Strategies and Styles of Migrant Female Domestic Workers in Singapore," *Urban Studies* 35, 3 (1998): 583–602.

16 Olivia Killias, "The Politics of Bondage in the Recruitment, Training, and Placement of Indonesian Migrant Domestic Workers," *Sociologus* 59, 2 (2009): 145–72; and Daromir Rudnyckyi, "Technologies of Servitude: Governmentality and Indonesian Transnational Labor Migration," *Anthropological Quarterly* 77, 3 (2004): 407–34.

17 HOME, annual report, April 2014 to March 2015, 7.

18 Ibid.

19 Ibid., 9.
20 HOME, "HOME Academy," https://www.home.org.sg/home-academy-1.
21 HOME, annual report, April 2014 to March 2015, 6.
22 Lenore Lyons and Yeong Chong Yee, "Migrant Rights in Singapore: Political Claims and Strategies in Human Rights Struggles in Singapore," *Critical Asian Studies* 41, 4 (2009): 586.
23 HOME, annual report, April 2014 to March 2015, 5.
24 Ibid., 10.

Acknowledgments

THE EDITORS ARE PROFOUNDLY grateful to the enormous transnational team that made this book possible. We would like to thank our collaborators at the National Film Board of Canada (NFB), without whom this project would never have taken shape. *Highrise* was the brain child of director Katerina Cizek, who brought us all together. For research, photography, ideas, and inspiration, we thank Kate Vollum, Paramita Nath, Maria-Saroja Ponnambalam, Kristyna Balaban, Branden Bratuhin, Gerry Flahive, David Oppenheim, and Cass Gardiner. Where interviews are used, the authors have obtained permissions from interview participants.

As the home to most of our team and the site of substantial work together, there are many people to thank in Toronto, many of whom were also members of the NFB team. We are grateful to the 2667–2677 Kipling Avenue Tenants' Association members and community researchers: Obeng-Asare OB Kwabena, Faith Senior, Donna Bridgeman, Cheryl Turner, Jaymini Gina, Maria Gelardi, Nahatil Corvil, Hadi Hamza, Pritvanti Patel, Rita Murad, Nasra Shammo, Eleanor Jiminez, Sandra Osaradion, Romta Shlaymoon, Nupur Patel, Gloria Ehiogie, Sadaf Saadat. We also thank the research and documentary team: Heather Frise, Jordana Wright, Kristyna Balaban, Maria-Saroja Ponnambalam, Lauren Ash, Chris Romeike, Adam Makarenko, Igal Petel, Mike Myrden, Sean Feldstein, Jamie Hogge, Maya Bankovic. We are grateful to Russell Mitchell from Action for Neighbourhood Change and Microskills, Elise Hug and Eleanor McAteer from the City of Toronto's Tower Renewal office. We thank Graeme Stewart of ERA architects, Helios Design Labs, and Secret Location.

In Mumbai we had the opportunity to learn from community members and organizers from many corners of the city. We are very grateful to Anita Patil-Deshmukh and the team at PUKAR, to the residents of Campa Cola, Gaurav Enclave, Maya Nagar, and Chandresh Terrace, and to the activists, scholars, journalists, and filmmakers who took time to meet with us and share their work – including Swapna Bannerji Guha, Surabhi Sharma, Chandrashekhar Prabhu, YUVA, Meena Menon, Mukesh Patil, Miloon Kothari, and URBZ. Special thanks to Deepti and Mithu Ghoshal for showing our team incredible and intimate hospitality. We were also fortunate to work with students from KRVIA – Shreeamey Phadnis, Apoorva Deshpande, Prerna Shetty, Purva Dewoolkar. We are grateful to Hussain Indorewala and Shweta Wagh for the connection, and for orienting out team to this extraordinary city.

In Singapore we were fortunate to work with many extraordinary people and organizations. We are particularly grateful to all the migrant domestic and construction workers who spent time with us, and shared their stories, and their spaces. Thank you to TWC2, and especially to Noor Abdul Rahman, Debbie Fordyce, Shelley Thio, N. Karno, and Alex Au. Our deep thanks to HOME, especially Jolovan Wham. Thank you to Peck Hoon and Kristian from the stoptrafficking campaign, and to Healthserve, Singapore.

A Partnership Development Grant from the Social Science and Humanities Research Council of Canada was critical in making this research possible. For generous support that contributed to the publication of this book, we are grateful to the Pierre Elliott Trudeau Foundation.

Finally, we would like to thank all the authors who contributed to this unusual collection, for sharing their brilliance and for their patience with our long production process, as well as the staff at UBC press, for their exemplary support in bringing this book into being. We especially thank James MacNevin for his commitment to this project and his endurance throughout the process, and Michelle van der Merwe for surviving the long and tangled work of helping this project come together.

Contributors

GRACE BAEY is a photographer and filmmaker with an interest in issues of gender, identity, and labour migration in Southeast Asia. She was previously a research associate at the Asia Research Institute, National University of Singapore.

SIMONE BROWNE is an associate professor in the Department of African and African Diaspora Studies, University of Texas at Austin.

CHARMAINE CHUA is an assistant professor of Global Studies at the University of California, Santa Barbara. Her work has been published in *EPD: Society and Space, Historical Materialism, and Political Geography,* and several edited volumes.

KATERINA CIZEK is a Peabody and Emmy-winning documentary director working in emergent digital forms. She leads the Co-Creation Studio at the Massachusetts Institute of Technology's Open Documentary Lab.

DEBORAH COWEN is a professor in the Department of Geography and Planning at the University of Toronto. She is the author of *The Deadly Life of Logistics: Mapping Violence in Global Trade, Military Workfare: The Soldier and Social Citizenship in Canada;* and coeditor of *War, Citizenship, Territory* and the *Errantries* book series at Duke University Press. Deborah serves on the board of the Groundswell Community Justice Trust Fund and was a collaborator on the National Film Board of Canada's Emmy award-winning *Highrise* project.

JUDY DUNCAN is the head organizer of ACORN Canada, which she started in 2004.

NEHAL EL-HADI is a writer, researcher, and editor whose work explores the intersections of, and interactions between, the body (racialized, gendered), place (urban, virtual), and technology (internet, health).

HEATHER FRISE is a multidisciplinary artist and filmmaker. She teaches at the Ontario College of Art and Design University.

STEPHEN GRAHAM is a professor in the School of Architecture, Planning and Landscape, Newcastle University.

JU HUI JUDY HAN is assistant professor in Gender Studies at the University of California, Los Angeles. She writes and teaches on political cultures, religions and secularisms, and (im)mobilities.

HUSSAIN INDOREWALA and **SHWETA WAGH** are activists, architects, and scholars working in Mumbai. They are involved in a wide range of community organizing and participatory planning initiatives in the city.

SYMON JAMES-WILSON is a graduate of the Department of Geography and Planning at the University of Toronto. Her work explores questions of mobility, infrastructure, race, and critical geographies of education.

ANJA KANNGIESER is a political geographer and sound artist holding a Vice Chancellor's Research Fellowship in Geography, University of Wollongong, Australia.

SAMEERA KHAN is an independent journalist and researcher, and visiting faculty in journalism at the School of Media and Cultural Studies at the Tata Institute of Social Sciences, Mumbai. She is a coauthor (along with Shilpa Phadke and Shilpa Ranade) of the critically acclaimed book *Why Loiter? Women & Risk on Mumbai Streets*.

JAMES KILGORE is an activist and co-director of the FirstFollowers Reentry Program. He has written widely on mass incarceration and e-carceration, including *Understanding Mass Incarceration: A People's Guide to the Key Civil Rights Struggle of Our Time*, and four novels, all of which were drafted during his six-and-a-half years in prison.

KRYSTLE MAKI specializes in community based research and applied sociology and currently works at Women's Shelters Canada as research and policy manager.

SHAKA McGLOTTEN is professor of Media Studies and Anthropology at Purchase College-SUNY.

ALEXIS MITCHELL is an artist and scholar with a doctorate from the University of Toronto. She has had recent exhibitions at Mercer Union (Toronto), the Berlinale (Berlin), and the BFI London Film Festival, and was an artist-in-residence at Akademie Schloss Solitude, Stuttgart, in 2015–17 and at the MacDowell Colony in New

Hampshire in 2018. She often works collaboratively with artist Sharlene Bamboat under the name Bambitchell.

LIZE MOGEL is an interdisciplinary artist whose work has been shown at the 2006 Gwangju Bienniale (South Korea), PS122 (NYC), Eyebeam (NYC), and others.

PARAMITA NATH is an independent documentary maker based in Toronto and New York. She works in both traditional and emerging platforms, experimenting with new approaches to storytelling.

NATALIE OSWIN is associate professor in the Department of Human Geography at University of Toronto Scarborough, managing editor of the interdisciplinary journal *Environment and Planning D: Society and Space*, and author of *Global City Futures: Desire and Development in Singapore*.

EMILY PARADIS is an instructor with the Urban Studies Program of Innis College at University of Toronto, a Maytree fellow, a collaborator with the Canadian Observatory on Homelessness, a member of the Right to Housing Coalition, and an independent research consultant. She has authored more than thirty publications on housing policy, homelessness, human rights, and lived expert leadership.

SHILPA PHADKE is a sociologist and assistant professor at the School of Media and Cultural Studies at the Tata Institute of Social Sciences, Mumbai. She is a coauthor (along with Sameera Khan and Shilpa Ranade) of the critically acclaimed book *Why Loiter? Women & Risk on Mumbai Streets*.

SASKIA SASSEN is the Robert S. Lynd Professor of Sociology and is the former cochair of the Committee on Global Thought at Columbia University.

R. JOSHUA SCANNELL is an assistant professor of Digital Media Theory in New School's School of Media Studies.

KASHAF SIDDIQUE is a youth resident of Chandresh Terrace in Mira Road, a suburb of Mumbai, who became active in organizing in response to the attempted demolition of her home.

NICOLE STAROSIELSKI is associate professor in the Department of Media, Culture, and Communication at New York University. She is author of *The Undersea Network*, and coeditor of *Signal Traffic: Critical Studies of Media Infrastructure*, *Sustainable Media: Critical Approaches to Media and Environment*, and the *Elements* book series at Duke University Press.

BRETT STORY is an assistant professor in the School of Image Arts at Ryerson

University, has a PhD in geography from the University of Toronto, and has held fellowships from the Guggenheim Foundation and the Sundance Documentary Institute. She is the author of *Prison Land: Mapping Carceral Power across Neoliberal America* and the director of the award-winning documentaries, *The Prison in Twelve Landscapes,* and *The Hottest August.*

INDU VASHIST is the executive director of the South Asian Visual Arts Centre, as well as a yoga teacher, and independent writer and researcher.

VISUALIZING IMPACT creates data-driven tools to advance a factual, rights-based narrative of the Palestinian-Israeli issue.

ALAN WALKS is a professor of urban geography and planning at the University of Toronto. He has published numerous scholarly articles examining urban inequality from different perspectives, is the editor of *The Political Economy and Ecology of Automobility: Driving Cities, Driving Inequality, Driving Politics,* and the coeditor of *The Political Ecology of the Metropolis.*

Index

Note: Page numbers with (f) refer to illustrations.

France: freedom of expression, 66; national broadband plan, 67
Frise, Heather, 12, 21–45, 194–99, 274
Fuller, Buckminster, 91
fusion centres, 86, 92, 93*n*10, 94

Gaurav Enclave. *See* Mumbai, illegal buildings, Mira Road (Gaurav Enclave)
Geiger, Susan, 9
Gelle, Hibaq, 73, 75, 76
gender. *See* domestic workers; right to the city; Singapore, migrant construction workers; Singapore, migrant domestic workers; women
gentrification, 26, 110, 215, 219
Germany: censorship, 188–89; electronic monitoring (EM), 108–9; internet as legal right, 66
Ghana: do-it-yourself Wi-Fi, 196; transnational lives, 34
Ghertner, Asher, 147
Girls at Dhabas (women in tea shops), 170–71
global financial crisis (2008): Canada, 25, 46, 48–49, 52–54, 53(f); United States, 38, 48–49
global infrastructure: submarine cables, 12, 200–205, 201(f)–4(f); telegraph networks, 200–201
globalized digital life: about, xvi–xvii, 5–9, 13–15; activism, 4–7; digital citizenship, xvi–xvii, 5–6, 34–35, 138–39; economic inequalities, 4–5; hierarchies of groups, 5–6, 9; power relations, 5–6, 9; public vs private networks, 138–39; revolts (2011), 4–6, 8; urban citizenship, 5–9, 34–35; urbanization, 6–7, 14–15; vertical lives, xiii. *See also* access to digital technologies; activism; borders, digital; digital divides; *Highrise* projects (NFB); migrant labour; Mumbai; policing; security; Singapore; social media; surveillance; Toronto; transnational lives
Goh, Daniel, 215, 217, 228

Gonzalez, Joaquin, 213
GPS (Global Positioning System): app-based taxi services (India), 182; electronic monitoring (EM), 108–9; surveillance of domestic workers, 263
Graham, Scott, 26
Graham, Stephen, 11, 85–88, 274
Greece: do-it-yourself Wi-Fi, 194–97
Greene, Daniel, 214–15
Gross, Aeyal, 186
growth in neoliberal planning, 150
guest workers. *See* migrant labour
Guifi network, 195

Han, Ju Hui Judy, 12, 175–78, 274
Harris, Nigel, 148–49
Harvey, David, 214
Hawes, Emily, 26
HDB (Housing Development Board) gallery/museum, Singapore, 209–10, 227, 229*n*1. *See also* Singapore, housing
Herbert, Steve, 110
Hermer, Joe, 40
heteronormative space in Singapore, 249, 252, 258*n*4. *See also* Singapore, everyday life of citizens
Highrise projects (NFB): about, ix–xii, xv–xvii, 3–4, 9–10; awards, x; cocreation methods, ix–x, xiii–xiv, 9–10; digital citizenship, xvi–xvii; funding, x–xi; *Highrise*, ix–xiv, 3–4, 44*nn*32–33, 142*n*14, 142*n*17, 142*n*19, 194; limitations, x–xi; localities and urban spaces, xv–xvii; Mumbai site, 8–10, 11–12, 142*n*14, 142*n*17, 142*n*19; production technologies, xiii; revolts (2011), 4–6, 8; situated solidarities, 9–10; surveys, xi–xii, 28–29; Toronto site, x–xii, 7–13, 24–25, 28–29, 44*nn*32–33; *Universe Within*, x, xiii, 4, 142*n*14, 142*n*17, 142*n*19
high-rises. *See* Singapore, housing; Toronto, inner suburbia high-rises; Toronto, Rexdale high-rises
Hinduism, 33, 33(f)

The Holocaust's Visit to Yad Veshem (video, Cohen Vaxberg), 190–91
HOME Academy, education for domestic workers, 219, 222, 264–67
Hong Gi-tak, 178
Hong Kong: submarine cables, 12, 200–205, 203(f)
hooks, bell, 74
Horn, Martin, 110–11
housing. *See* financialization; Mumbai, financialization and real estate; Mumbai, illegal buildings; Singapore, housing; Singapore, migrant labour; Toronto, inner suburbia high-rises; Toronto, Rexdale high-rises
How Would You Manage without the Holocaust? (video, Cohen Vaxberg), 191
Huang, Shirlena, 213
human rights: digital technologies as, 40, 65–66; electronic monitoring (EM), 108–9; freedom of expression, 66; surveillance issues, 108–9; UN right to the internet, 65–66, 194. *See also* right to the city

illegal buildings. *See* Mumbai, illegal buildings; Mumbai, illegal buildings, Campa Cola; Mumbai, illegal buildings, Mira Road (Chandresh Terrace); Mumbai, illegal buildings, Mira Road (Gaurav Enclave)
incarceration. *See* prisons
India: access to digital technologies, 180–81; app-based taxi services, 179–85; digitization and financialization, 138; foreign direct investment (FDI), 136–37, 181–82; haggling for services, 182–83; Indigenous peoples, 119, 141*n*10, 153; internet use, 180–81; labour unions, 184; migrant labour, 153; national urban renewal, 137; planning keywords, 148–51; rent control, 143; Right to Information (RTI), 142*n*15, 160; sand industry, 153; slum redevelop-

ments, 120, 129, 147–48; smartphone use, 180–81; special economic zones, 137; taxi industry, 179–85. 167; transnational domestic worker networks, 265; Urban Land Ceiling Act, 119, 141*n*9; wallet apps, 182; women in tea shops (Girls at Dhabas), 170–71; women's street safety (#WhyLoiter), 167–72; work ethic, 183–84. *See also* Mumbai
Indonesia: maritime boundaries, 239; migrant labour in Singapore, 226–27, 244; sand exports, 13, 240, 244; transnational domestic worker networks, 263–65
Indorewala, Hussain, 12, 136, 145–52, 274
infographics on digital divide, 112–13, 113(f)
international digital life. *See* global infrastructure; globalized digital life; transnational lives
international law and declarations: freedom of expression, 66; law of the sea, 241; maritime boundaries, 241; right to the internet, 65–66, 194; sound warfare, 103
Iraq, transnational lives, 32
Israel: biopolitical management, 191; Cohen Vaxberg's online protests, 186–93; criminal charges for "defiling national symbols," 186; smartphone use, 189; Visualizing Impact, 11, 112–13, 113(f)

Jamaica, transnational lives, 34
James-Wilson, Symon, 13, 224, 260–69, 274
Japan: submarine cables, 12, 200–205, 202(f)
Jayaprakash, Abhishek, 168
Jinsook Kim, 176
John, Nishant, 168
Jones, Feminista, 80
Jordan, June, 90–93
Joseph, Daniel, 214–15
justice. *See* activism

(#WhyLoiter), 167–72, 168(f), 171(f), 172(f)

Mumbai, digital technologies: about, 118, 132–34, 137–41; activism, 124–27, 126(f), 132–34, 140–41, 142*n*20; community-building role, 140–41; ICTs, 124–27; international news, 127, 154; limitations of, 132–33; social media, 124, 127, 132–34, 154

Mumbai, financialization and real estate: about, 117–18, 135–37, 163; corruption, 120, 124, 126–29, 134, 142*n*12, 153, 156–57, 160–61; debt, 128, 154–55; demolitions, 118, 137; development régimes, 12, 130–31, 136–38, 142*n*15, 143*nn*21–22, 145–48, 153; digital infrastructure for, 134, 138; displacements, 117–18, 134–35, 147–48, 163; dispossession through debt, 128, 135–36, 140, 154–55; FDI (foreign direct investment), 136–37; FSI (floor space index), 136, 145–47, 164; historical background, 135–37, 145; keywords for neoliberal planning, 148–51; migrant labour, 153; mortgage markets, 117; neoliberalism, 131–32, 146–51; ownership of building units, not land, 128–29; real estate industry, 117, 135–40; rent control, 143*n*21; slum redevelopment, 134–37, 147–48; special economic zones, 137; speculation, 118–19, 128, 135–38, 143*n*22, 146, 153, 163; timeline, 114–15

Mumbai, illegal buildings: about, 117–19, 129–30, 135; activism, 124–27; building collapses, 121, 165–66; city services, 119; corruption, 124, 127, 128–29, 142*n*12; debt, 128, 154–55; demolitions, 12, 117–18, 128, 135; development régimes, 12, 130–31, 142*n*15; dispossession through debt, 128, 135–36, 140, 154–55; illegality of building, 135; legal actions, 129–31; occupancy certificates, 119, 121, 130–31, 141*n*7; ownership of units,

not land, 128–29, 135; property rights, 118, 128, 129–30, 135, 143*n*24; social class, 10, 131–35; statistics, 117, 142*n*12, 154, 160, 163; substandard construction, 129, 162(f), 164(f); timeline, 114–15

Mumbai, illegal buildings, Campa Cola: about, 114–15, 118–19, 160–66; activism, 124–27, 132–34, 142*n*20, 161; allowable density, 119; city services, 120; corruption, 129, 160–61; demolition, 119, 124, 132, 160–64, 161(f), 165(f), 166(f); development régimes, 12, 130–31, 142*n*15, 164–66; evictions, 124, 160, 165; FSI (floor space index), 164; historical background, 118–21, 127–28; illegal buildings, 119, 129–30, 164–66; lawsuits, 130–31, 160, 165; media coverage, 160–61; occupancy certificates, 121, 130–31, 141*n*7; property rights, 129–30, 143*n*24; social class, 124, 131–34, 160–61; social media, 10, 124, 132–34, 154, 161, 161(f), 162(f); speculation, 119; substandard construction, 129, 162(f), 164(f); timeline, 114–15; water Mafia, 120

Mumbai, illegal buildings, Mira Road (Chandresh Terrace): about, 114–15, 117–18, 153–59, 155(f); activism, 124–27, 126(f), 133–34, 142*n*20, 154–59; affordable housing, 120; building society, 122, 124, 156–59; city services, 120–21; commutes, 132, 140, 142*n*14, 153; corruption, 120, 124, 127, 128–29, 134, 156–57; demographics of residents, 120, 122, 154, 157–58; demolition, 12, 122–24, 123(f), 125(f), 127–28, 132, 154–59; development régimes, 12, 130–31, 142*n*15; digital technologies, 127, 156; historical background, 118–21, 129, 157–58; illegal buildings, 129–30, 142*n*12; illiteracy, 157–59; lack of community, 131–32, 140–41; lawsuits, 130, 157–59; MBMC (Mira-Bhayander Municipal Council), 119–20, 124, 129;

occupancy certificates, 121, 130–31, 141*n*7; property rights, 129–30, 143*n*24; relocations from, 124; social class, 10, 127, 131–32; social media, 124, 127, 133–34, 154–59; story of Kashaf's family, 153–59; substandard buildings, 120, 123(f), 124(f), 129, 139(f), 155, 155(f); timeline, 114–15

Mumbai, illegal buildings, Mira Road (Gaurav Enclave): about, 114–15, 120–24, 121(f), 122(f), 153–54; activism, 124–27, 133, 142*n*20; Building Society, 121, 124; city services, 120–21; commutes, 132, 140, 142*n*14, 153; corruption, 120, 124, 126, 128–29; demolition, 129, 132, 154; development régimes, 12, 130–31, 142*n*15; digital infrastructure, 120–21, 127; evictions, 121, 126, 128, 133; historical background, 118–21; illegal buildings, 129–30; lack of community, 131–32, 140–41; occupancy certificates, 121, 130–31, 141*n*7; ownership of unit, not land, 128–29; property rights, 129–30, 143*n*24; social class, 120–21, 131; social media, 124, 126, 132–33, 154; substandard construction, 120, 121(f), 122(f), 126, 129, 139(f); timeline, 114–15

Myanmar: migrant labour in Singapore, 226–27, 244; sand exports, 13, 240, 244, 246*n*4; transnational domestic worker networks, 265

Nagar, Richa, 9
Nath, Anjali, 90
Nath, Paramita, 12, 160–66
neoliberalism: debt and risk transfer, 27; digital debt in neoliberal economy, 39–40; frameworks, 149–50; growth, 150; keywords in planning, 148–51; partnership/participation, 150–51; planning keywords, 148–51; policing partnerships, 98, 100; real estate speculation, 27; simplification, 150; vision,

149; welfare state dismantling, 40, 57, 61–62. *See also* Mumbai, financialization and real estate; Ontario Works (work-for-welfare)

network cables, submarine, 12, 200–205, 201(f)–4(f)

New America Foundation, 196

New York City, policing: about, 94–101; broadband infrastructure, 97; CCTVs, 95; Domain Awareness System (DAS), 11, 95–100; heat maps, 96–97; predictive policing, 11, 92, 94–95, 96–100; racialized people, 91–92, 95; urban security, 85–88. *See also* policing

New Zealand: submarine cables, 12, 200–205, 203(f)

Nigeria, transnational lives, 34

Noble, Safiya Umoja, 39

Norway: LRAD sonic policing, 103

ocean/land boundaries. *See* borders, coastal

Oh Soo-young, 176

Ola Cabs, Chennai, 180–84

Ontario. *See* Canada; Toronto

Only Death Will Discharge Us from the Ranks (Cohen Vaxberg), 190

Ontario Works (work-for-welfare): about, 11, 57–62; anti-fraud measures, 57, 61; documentation, 59–61, 60(f); maintenance enforcement, 58(f), 59; moral discourses, 11, 40, 61; neoliberalism, 40, 57; privacy rights, 59–61, 60(f); surveillance system, 11, 57–61, 58(f), 60(f)

Open Technology Institute, 196

Oswin, Natalie, 13, 248–59, 275

Pak Jun-ho, 178

Pakistan: Girls at Dhabas (women in tea shops), 170–71

Palestine: Cohen Vaxberg's online protests, 186–93; Israel's untreated wastewater, 191; Visualizing Impact, 11, 112–13, 113(f)

workers; Singapore, migrant
domestic workers

Singapore, migrant construction workers:
about, 211–15, 216(f), 233–37; activism,
225; Bangladeshi workers, 217, 226,
233–37, 244; dorms, 213, 217, 219–20,
228, 256(f); fees to access work, 212,
226, 234; Little India, 216(f), 217–
19, 218(f), 220(f); nationalities, 211,
213, 244–45; terraforming projects, 210,
239–45; training, 213, 234; wages, 213,
234; workplace injuries, 212, 224–25,
226, 235–36, 236(f), 245

Singapore, migrant domestic workers:
about, 14–15, 211–13, 220–23; access to
digital technologies, 221–23; activism,
225–26; demographics, 211, 262; education, 225–26, 263–67; HOME (vocational training), 219, 222, 264–67;
immigration policy, 214, 223, 225, 256–
57; live-in workers, 222–23; nationalities, 211, 221, 225, 226; personal and
community networks, 225–26, 262,
264–67; social media, 225, 262–63; support groups, 212(f), 226, 227(f), 266–67;
surveillance, 14–15, 221–23, 228–29;
traditional communications, 264–67;
transnational lives, 224–25, 262–63;
TWC2 (advocacy group), 212(f), 213,
217, 220, 223, 226, 230; wages, 213–14;
work conditions, 213, 221–23, 263

Singapore, territorial expansion: about,
238–47; corruption, 240–41; environmental impacts, 244–45, 246n4; historical background, 239–40, 243–44; map,
240(f); maritime boundaries, 239; port
development, 238; as "reclamation,"
242–45; sand imports, 13, 240–41,
246n4; terraforming, 210, 239–45

Singh, Neha, 168

slum redevelopment, Mumbai, 134–37,
154, 163, 166. *See also* Mumbai

smart cities, 94

Smith, Andrea, 77–78

Snapchat. *See* social media

social activism. *See* activism

social assistance, Ontario. *See* Ontario
Works (work-for-welfare)

social media: about, 4–7; digital divide,
171–72; for documentation of corruption, 155–56; impact on activism, 4–7,
124, 132–34, 161(f) 162(f), 167–68; impact on everyday life, 184–85; limitations of, 132–34, 262–63; news reports,
154–55, 161; public vs private networks,
138–39, 196–97; revolts (2011), 4–6, 196;
right to the city, 167–72; for surveillance
of domestic workers, 222; training in,
225. *See also* access to digital technologies; activism; transnational lives

Song, Stephen, 195–96

sonic governance of space, 11, 102–7

South Africa: national broadband plans, 67

South Korea: high-altitude protests, 12,
173–78, 177(f); Listen to the City, 176,
177(f); smart cities, 94

spatial fix, 214–15, 227

speech recognition, 104–6

Starosielski, Nicole, 12, 200–205, 275

Story, Brett, 3–17, 117–44, 275–76

submarine cables, 12, 200–205, 201(f)–4(f)

surveillance: about, 11, 85–88; biometrics,
86, 219–20; DAS system in NYC, 95–96;
electronic monitoring (EM), 11, 105–6,
108–9; by FBI in US, 89–93; fusion centres, 86, 92, 93n10, 94; GPS systems,
108; human rights, 108–9; monitoring
technology, 108–11; multiple systems,
86–87; terrorist threats, 96; voice recognition, 104–6; work-for-welfare programs, 11, 57–61, 58(f), 60(f). *See also*
CCTV (closed-circuit television
cameras)

Sweden: electronic monitoring (EM), 108

Syria, transnational lives, 32

Tan, Kenneth Paul, 254

taxi industry, India, 167, 179–85

telegraph networks, 200–201

temporary migrant workers. *See* domestic workers; migrant labour; Singapore, migrant labour; Singapore, migrant construction workers; Singapore, migrant domestic workers

territorial expansion. *See* Singapore, territorial expansion

textual assemblages, 11, 73–81

Thio, Shelley, 213, 221, 223

Thiyagarajah, nayani, 74

Thompson, Clive, 197

The Thousandth Tower (photo-blog), xi, 43–44*n*28

time as numbers in high-altitude protests, 12, 173–78, 177(f)

Toronto: about, 3–4, 22–25; demographics, 25, 27, 42*n*9; digital technologies, 28–32; immigrants, 24–25, 27–28, 32; inner city, 4, 24–26; inner suburbs, 3–4, 24–25; neoliberal economy, 27–28; planetary urbanization, 6–7; right to the city, 73–81; social programs, 27–28; transit, 29, 32; work-for-welfare programs, 57–62; workplaces and digital technologies, 31–32. *See also* Canada; Canada, digital technologies; Ontario Works (work-for-welfare); Toronto, inner suburbia high-rises; Toronto, Rexdale high-rises

Toronto, debt: about, 11, 25–28, 46–56; city core, 26; costs and contracts for digital services, 35–39, 46, 70; debt-to-income ratios, 25–26, 46, 49–54, 51(f), 53(f); demographics, 49–51, 51(f), 54; digital debt, 22–23; disposable income, 49; financialization, 47–48; global financial crisis (2008), 25, 46, 52–54, 53(f); home equity lines of credit, 48, 49; income inequality, 46; low wages, 36, 47; mapping of, 49–51, 51(f); mortgages, 25, 48; payday loans, 38–39, 49, 70; unsecured debt, 25–26, 49–52, 51(f), 52–54, 53(f)

Toronto, inner suburbia high-rises: about, 3, 7–13, 22–25, 23(f)–24(f); demograph-

ics, 25, 27, 32, 42*n*9; living conditions, 3, 24, 35; tenants, 24–25; tenants' associations, 22, 29, 41, 41(f), 44*n*29; tenants' debt, 22, 50

Toronto, Rexdale high-rises: about, 3, 22–25, 23(f)–24(f); demographics, xi–xii, 25, 27, 32, 42*n*9; *Highrise* projects, x, xi–xii, 7–13, 24–25, 28–29; living conditions, 3, 24, 35; local improvement projects, 29, 44*n*29; surveys, xi–xii, 28–29; tenants, 24–25, 29, 32; tenants' associations, 22, 29, 41, 41(f), 44*n*29; transit, 8, 29, 32; urban citizenship, 5–9, 34–35. *See also Highrise* projects (NFB); Toronto, inner suburbia high-rises

Toronto, Rexdale high-rises, digital technologies: about, 29–41; access to services, xii, 7–8, 29, 31, 40; activism, 40–41, 41(f); costs and contracts, 22, 35–40, 42*n*2; digital debt, 7–8, 10–11; digital divides, 7–8, 34–35; education and entertainment, 30(f), 31, 34, 36; essential need for, 7–8, 10–11, 31–32, 40–41; gender and age differences, 31; as human right, 40; money transfers, 8; neoliberal economy, 39–40; patterns of use, 29; payday loans, 38–39, 49, 70; purposes, 29–35; religious practices, 31, 33, 33(f), 36(f); resistance to, 40–41; social media, 31, 32; spatial inequalities, 38–39, 42*n*9; transnational lives, 7–8, 32–34, 33(f); workplace ICTs, 31, 35

towers, residential. *See* Singapore, housing; Toronto, inner suburbia high-rises; Toronto, Rexdale high-rises

traditional communication technologies, 264–67

transnational lives: about, 32–35, 33(f); digital divides, 35; domestic worker networks, 265; nostalgia and grief, 34; online dating, 225; "online vs real life," 75; transnational mothering, 15, 224, 262–63. *See also* globalized digital life; social media